T0134638

The Contribution of Technology to Added Value

António S. C. Fernandes

The Contribution of Technology to Added Value

António S. C. Fernandes
Instituto Superior Técnico, Taguspark
University of Lisbon
Porto Salvo
Portugal

ISBN 978-1-4471-5863-9 ISBN 978-1-4471-5001-5 (eBook)
DOI 10.1007/978-1-4471-5001-5
Springer London Heidelberg New York Dordrecht

© Springer-Verlag London 2013
Softcover re-print of the Hardcover 1st edition 2013
This work is subject to copyright. All rights are reserved by the Publisher, whether the whole or part of the material is concerned, specifically the rights of translation, reprinting, reuse of illustrations, recitation, broadcasting, reproduction on microfilms or in any other physical way, and transmission or information storage and retrieval, electronic adaptation, computer software, or by similar or dissimilar methodology now known or hereafter developed. Exempted from this legal reservation are brief excerpts in connection with reviews or scholarly analysis or material supplied specifically for the purpose of being entered and executed on a computer system, for exclusive use by the purchaser of the work. Duplication of this publication or parts thereof is permitted only under the provisions of the Copyright Law of the Publisher's location, in its current version, and permission for use must always be obtained from Springer. Permissions for use may be obtained through RightsLink at the Copyright Clearance Center. Violations are liable to prosecution under the respective Copyright Law.
The use of general descriptive names, registered names, trademarks, service marks, etc. in this publication does not imply, even in the absence of a specific statement, that such names are exempt from the relevant protective laws and regulations and therefore free for general use.
While the advice and information in this book are believed to be true and accurate at the date of publication, neither the authors nor the editors nor the publisher can accept any legal responsibility for any errors or omissions that may be made. The publisher makes no warranty, express or implied, with respect to the material contained herein.

Printed on acid-free paper

Springer is part of Springer Science+Business Media (www.springer.com)

Contents

Symbols and Abbreviations

A	Production factor linked to technical progress, which appears in neo-classic equations of growth. In one of the possible formulations, A has the dimensions of productivity, i.e., [uv (units of value) per worker]. In models proportional to capital, Y = A.C, it is a dimensionless constant.
â	Growth rate of parameter A, $\hat{a} = \dot{A}/A$, $[ut]^{-1}$
C	Capital = TA + CA, [uv]
C̲	Value added by the use of capital assets, [uv]
CA	Capital assets, [uv]
C_f	Final consumption–SEC95, 3.74, $C_f = C_{pr} + C_{pu,}$ [uv]
CI	Capital Index, CI = C̲/GVA, dimensionless, [nd]
C_N	C_N = C/N Capital per worker, [uv/nw]
CS	Capital services (OECD definition), hours worked, [hw]
fe	Empiral correction factor, [nd]
GDP	Gross Domestic Product, [uv]
GVA	Gross Value Added, GVA = Y = L + T + C̲, [uv]
I	Investment, [uv]
ka	Technology coefficient, [nd]
KI	Knowledge Index KI = L/GVA, [nd]
kp	GVA coefficient, [nd]
L	Value added by the use of knowledge, [uv]
ln	Natural logarithm
LP	LP = Y/N Labour productivity, [uv/nw]
N	Number of workers or number of hours worked, [nw] or [hw]
NPV	Net Present Value, [uv]
POA	Profits on Ordinary Activity, [uv]
S	National savings (SEC95, 8.96), [uv]
s	Saving coefficient, s = S/Y, [nd]
t	Time, [s]
T	Value added by the use of technology assets, [uv]
TA	Technology assets, [uv]
TCP	Technological Content of a Product, [uv]
TI	Technology Index TI = T/GVA, [nd]
IPV	Intermediate Products Value, [uv]

ULC Unit Labour Cost, ULC = KI, [nd]
w Wage, [uv/N]
Y Product of the economy (normally GVA), [uv]
Y_N Product of the economy per worker, [uv/nw]
$\dot{Y}$ $\dot{Y} = \frac{\partial Y}{\partial t}$
α Dimensionless constant used in Cobb-Douglas production functions
δ Depreciation coefficient, [nd]
δa Average depreciation coefficient for technological forms, [nd]
δb Average depreciation coefficient for capital forms, [nd]

Chapter 1
Introduction

There is a wide consensus that technology contributes to create economic value, yet that process is not well understood and so technology's contribution to growth has never been assessed objectively. Why? There are ambiguous and controversial issues about what is technology.

This book proposes a conceptual framework that will allow dealing with the concepts of knowledge, technology, and capital as autonomous operational concepts. Its intended audience are managers, economists, and engineers, either in academia, or solving everyday problems in industry and services. First, it clears up the semantics, which is fundamental in any communication system. Moreover, it provides understanding a significant taxonomy for technology dependence and allows and modeling of how knowledge, technology, and capital individually contribute to production and to value adding.

1.1 Issues and Goals

There is a semantic problem with the concept of technology, in other words, there are different perceptions of its meaning. Being a new word, induced from the older word technique, the word technology was introduced to extend the meaning of technique to a wider field. However, the difference between the two meanings is not at all clear, and above all is not consensual. Specifically, management, engineering, and economy understand technology in three particular ways. Putting it in the simplest form, managers know they cannot run a process competitively without continuous technological innovation; engineers consider technology to be what they produce; and economists, as the production factors they know best are labor and capital, say that technology is everything else that may contribute to value added. Additionally, finance and accounting completely ignore the concept of technology. Nevertheless, everyone agrees that it is fundamental for production and a most important factor for growth.

Semantic problems are best studied by philosophers tracing the concepts' epistemological evolution, an analysis concerning many sociological elements as well. Those

A. S. C. Fernandes, *The Contribution of Technology to Added Value*,
DOI: 10.1007/978-1-4471-5001-5_1, © Springer-Verlag London 2013

analyses have had important contributions from the sociology of sciences, but the results are not in a format on which objective production models could be built. In fact, the results show technology as a complex concept describing a web of skills, knowledge, and artifacts interweaving the whole of society. Apparently, they say technology is the structure we all leave with and within. Changes in technology will change society's structure; it may develop it to unimaginable ways or quickly destroy it. It is part of us, humans in society. This is thorough, interesting, most probably true but not operational.

Additionally, there is another problem which is directly related to some production models that use the concept of technology: a problem that leads to dubious conclusions. It is related to a basic scientific rule, which states that whenever we apply a mathematical identity to a tangible relation, the dimensions of the identity's left term must equal the dimensions of its right term. It is a fact that many mathematicians do not worry about dimensions when dealing with algebra and calculus. This happens because mathematics is an auto-correcting structure, meaning that errors and omissions are quickly spotted internally without the need to apply the analysis to the real world. Mathematical constructs, such as multidimensional analysis and tensor calculus, were developed with no clue as to what purpose to which they could be applied. Physics and engineering, however, are not self-correcting: we engineers have to be extremely careful when using mathematical functions, for instance, to model a cable-stayed bridge, write an information compression algorithm or apply the Einstein equation $E = m.c^2$. For energy (E) to be measured in Joule, and for the velocity square c^2 to be measured in square meter per square second, the parameter m must be measured in mass (kg). Also we cannot mix meters with inches or mass with weight. Nor in economics, where a Cobb-Douglas production function relating value or volume to technology, in which labor and capital must have, for the same reasons, identical dimensions on both sides of the equal sign. This basic, universal, and unavoidable rule has not always been attended to for the past 50 years, resulting in many questionable conclusions. This fact has also contributed to the current ambiguity surrounding the meanings of technology, in so far as it has been often represented by the parameter A, and A has been used with different dimensions, which makes comparisons doubtful.

These issues are the two immediate motives for the analysis presented in this book. In summary, the main goal is to contribute to a better understanding of the meaning of technology and proposing one way its role could be operationalized, building a model that would consider technology as a growth factor, assessed objectively and independently. In that way, it would be possible to compute the contribution of technology to that of added value.

1.2 Technology Versus Knowledge

After the World War II, economists and managers focused their attention on technology, and hence technology management was born. This new area triggered the research into cybernetics and later informatics and robotics even before the information age and the silicon revolution. Technology became the pulsing heart of the corporation and technological management the most pertinent area to reach high growth rates. Production productivity stretched new upswings especially in the USA and Japan. Along the next 20 years, the importance of both the individual knowledge contribution to technological development, and the whole knowledge base corporations were building upon, slowly became obvious. The paradigms of the knowledge creating company and the knowledgeable organisation are good examples. Technology management somehow gave way to knowledge management, which takes on an organic, almost human, and holistic view of the organization, promoting, and validating innovation by technology and developing strategic decisions.

Technology, technical progress, knowledge, and technological knowledge are terms that, in a number of contexts, have been and still are synonymous, clearly evidencing an overlapping of the concepts of technique, technology and knowledge. Furthermore, technology is often listed as an asset and so considered a form of capital. Such a vision makes it difficult to discriminate between technology and knowledge, let alone to use one as an independent and objective growth factor. However, in their essence, they have different meanings. A technique is a somewhat simpler succession of actions with a well-defined goal, while technology is more complex and so involves specific skills as well as material artifacts. Knowledge, finally, is in a stricter sense, mostly used to refer to intellectual information, which includes anything necessary to operate a technological artifact, for example. There must be a way of discriminating between them.

1.3 The Role of Technology

Historically, the sociological understanding of technology is predominant and has imposed its view on management, economy, and to some extent, on engineering. The word came from the minds of philosophers in the context of analyzing manufacturing management. Thus, it is above all a sociological concept. But sociological arguments are seldom objective and operational. For instance, they were used to entertain the post-modern illusion that technology might drive history as an autonomous causal agent of social change. Moreover, management, engineering, and economy need more down-to-earth concepts to deal with. Regardless, the word technology was easily and quickly adopted by management and engineering, even if differently, and eventually by economists who saw it as a good representation for the idea of technical progress.

The role of technology is understood by managers and engineers in a more prosaic manner. For engineering the matter is even simpler, because technology is the typical output of engineering work: design, soft, or hard products. They use their knowledge and skills and embody them in material, so that a specific function may be performed. This output is a product meant to have a specific role in the production process. It is a technological form, or technology. For a manager, technology is part of a firm's resources, just as people and assets, however and similarly to engineers, he is not sure how to distinguish between them.

The resource-based view of the firm, following traditional strategic business policy, started by considering assets and people and their unique specificities as the classical resources, and evolved throughout the last 25 years to the current multifaceted understanding of the technological and knowledge type of resources, from tools to human capital, including what they name as specific capabilities. This view takes resources and their particular endowments, to be leveraged by management in order to attain the firm's goals, achieving the competitive advantage that would assure the envisioned success. Technology, inimitable or not, is just one type of resource, and again is not objectively detachable from the lot. Also, that school considers knowledge as just another resource without special characteristics. However, technology and knowledge are not treated separately because management cannot objectively distinguish between them.

Innovation-driven growth in the context of free market economies is an almost consensual idea both in the mainstream and in evolutionary theories. The subject matter is wide and transversal but it can be traced back to Schumpeter and his "creative-destruction" principle. By innovation, it is technological innovation that is meant, which is, after all, the same notion as of technical progress. Neither Schumpeterian economics, whatever this means, nor evolutionary economics make a difference in the way in which the concept of technology and its role in production are understood.

Technology has been a keyword in the so-called national innovation systems, developed and monitored at the international level by OECD, where science and technology play the most important role and innovation is the chief concept [1]. The analysis of these systems targets the innovative performance of the knowledge-based economies by measuring all sorts of knowledge flows, such as industrial linkages, human resources, publications, patents, technological diffusion, etc. They are mainly concerned with private and public investment in R&D, stressing *that the flows of technology and information are key to the innovative process* (summary). The definition of these systems uses the words technology, knowledge, technological learning, skills, and artifacts as equivalent as far as the system is concerned. Knowledge, as embodied in human beings, is human capital but it is also embodied in technology. *Disembodied technology or knowledge also includes other know-how, patents, licences, trademarks, and software* [2]. Technological diffusion may be measured by purchases of machinery and equipment; and knowledge flows by either personnel mobility or the technology balance of payments (patents, R&D services, know-how, etc—OECD Glossary). Equipment is referred to as embodied technology, but also as technology, and technology as embodied knowledge, and so on. The OECD glossary defines technology as follows: *Technology refers to the state of knowledge concerning ways of converting resources into outputs*. In other technology related definitions, this state of knowledge can be processes, facilities, and methods of operation.

In summary, the current situation is that there is no way to understand the meaning of technology and its role in production and value adding without overlapping with the roles of human knowledge and capital.

1.4 Approach and Methodology

My view and initial hypothesis is that the intricate current social system and the complexity of what we now call technology, hinders a clear vision of the different roles played by human knowledge, technological forms, and capital. Also, I know that knowledge is a concept whose meaning is as old as humanity and surely can be addressed and understood without the need for such relatively new concepts as technology or capital. Thus technology, not merely a skill and being somehow embedded in material form, should be separable from knowledge: Firstly, knowledge is itself an autonomous concept; and secondly, because technology has a physical existence that is independent of individual humans. Furthermore, even if technology and capital could sometimes overlap, from a functional point of view, there must be sensible criteria that could separate them.

There is no need to distort, remake, or create new concepts using already existent words. On the other hand, I propose to use these words with more exacting meanings so that they can be disentangled.

For this purpose I propose the following methodology: To start by conducting epistemological analyses of the three concepts throughout history and covering all the relevant scientific fields. This examination will produce lists of central attributes and extensions for each of the three concepts, such that it will be possible to focus on the central attributes, and only then attempt to separate them clearly, meaning that each should have a central group of attributes different from the attributes of the others. Once this deconstruction is done, it will be necessary to reconstruct the three concepts accordingly to a specific criterion. This will result in three new operational concepts, maintaining their original central attributes with the added quality of being operational in a specific context, that is to say, they can be used as autonomous factors in an economic production model.

In summary, to relate technology to growth and measure its contribution to value added, one should redesign, reinterpret or reconstruct the concept of technology in a way that it becomes independent of knowledge and capital while remaining compatible with its present meaning. This reconstruction will be accomplished by taking advantage of the methodological difference between a concept and an operational concept and building, out of the current concepts, new operational concepts for the three ideas of technology, knowledge, and capital. This is the contextual approach and the basic methodology sustaining the analysis described in this book.

1.5 The Process of Value Adding

Value adding describes a society's production system, the output of which is everything we consume and its value is termed as the gross value added (GVA). The gross domestic product (GDP) of a national economy is equal to the GVA added to indirect taxes, like the value added tax (VAT). As such, a production system is mostly assessed by analyzing the process of value adding, in other words, how the total GVA is achieved. In a single economy, like a national economy, or in a specific economic activity sector, like the manufacturing industries, the total GVA is the sum of all GVA contributions from each economic activity unit, such as a firm. Therefore, analyzing the value adding process in one firm is sufficient an effort to understand the whole value adding process. This is done using a rather objective language, accounting, which corresponds to international standards and is structured as a self-correcting system, just like mathematics.

By definition, the GVA in a firm is the subtraction of the cost of materials and consumables as well as other operating taxes and charges, from the total operating income. This difference is equivalent to the sum of four contributions: Staff costs, value adjustments on non-financial assets, taxes on profit, and profits on the ordinary activity. This algorithm plus a number of fundamental identities are the pillars of all financial and accounting information about firms, sectors, and economies. All indexes and ratios, as well as all the growth accounting models are based in this accounting language and system. We will also use this system to assess and compute the technology contribution to value added.

1.6 How can Technology Contribution be Assessed

I propose a model to describe the value adding process, where GVA equals the sum of the value contributions of the use of three independent production factors: Knowledge, technology, and capital. In other words, my hypothesis is that the GVA originates in the use of knowledge, in the use of technology and in the use of capital. Of course this model presupposes that knowledge, technology, and capital are independent and autonomous factors, which can only be true if they are independent operational concepts.

Next, I compare the standard GVA accounting algorithm with the one I propose, verifying, for each account, how much can be interpreted as originating in either knowledge, technology, or capital. In this way, we can compute the value contributions to GVA of the use of each of the three factors.

1.7 How the Book is Structured

Chapter 2 is dedicated to growth models, where it is shown how the idea of technology has been interpreted and used so far. Classic and more modern models are explained, with particular focus on the difficulty their authors demonstrate when discerning technology. It is also demonstrated why some of the conclusions from these models are not trustworthy, namely total factor productivity analyses. Following that, a new linear model (KTC model) is proposed, building it qualitatively step by step, fully justifying why the GVA can be calculated from the contributions of the use of knowledge, technology, and capital.

Chapter 3 is dedicated to the KTC model. Firstly, the operational concepts of knowledge, technology, and capital are constructed while a summary of the deconstruction analyses is described and the full study placed in a final annex. Secondly, the final algorithm needed to compute the values added through the use of knowledge, technology, and capital is established. Concurrently, the knowledge index, technology index, and capital index are defined, expressing their relative contribution to GVA. This rationale is also used for deduction of the main economic growth conditions.

Chapter 4 shows the technology index values for different economic activity sectors in Portugal and for manufacturing sectors of several European countries, clearly expressing their technology dependence.

In Chap. 5, a full new technology dependence taxonomy is proposed based on a statistical analysis of the technology index. Furthermore, the OECD technology intensity factor currently in global use is explained, as well as how it compares with the technology index proposed here, showing why the latter reflects better the technology dependence. Finally, how this model allows the computing of the technological content of any product is explained, by adding the technology contributions along the value chain.

Chapter 6 explains the concept of value, its origins and different types, including economic value. Definitions of consumed, restored, and created value are proposed, computing for the latter, the value created or destroyed during the last decade by several European countries. The knowledge-value-knowledge cycle is explained, and consequently why human knowledge is the direct origin of the value concept and so how economic value reflects the knowledge contribution to production. It is concluded that value is the metric for assessing knowledge.

Chapter 7 draws the main conclusions of this investigation. Comparing the initial hypotheses and goals with the results, the usefulness of the proposed model and its yields is concluded.

A long annex is placed at the end, where full deconstruction analyses of the concepts of knowledge, technology, and capital are described as well as how the respective operational concepts are rebuilt.

References

1. OECD (1997) National innovation systems. Paris. http://www.oecd.org/science/innovationinscienctechnologyandindustry/2101733.pdf. Accessed Oct 2012
2. OECD Glossary. http://stats.oecd.org/glossary/search.asp. Accessed Oct 2012

Chapter 2
Technology in Growth Models

For about 100 years, the discipline of economic growth and its models use the idea of technology (technical progress) in a rather objective fashion, quantifying its contribution to value production. This discipline is a crucial area where it is possible to realize how the concept of technology has been understood so far. The main objective of this chapter is to describe several growth models such that the role of technology is described as captured by their authors and eventually quantified. It will be shown how ambiguously and inaccurately the concept of technology is used.

The issue is how does technology impact on a transformation process and thus on economic systems? There is a consensus that it does impact and many authors modeled economic systems in order to show how, but it is not straightforward to understand their conclusions, because the concept of technology itself has not been made clear. An equivalent issue is how can technology contribute to growth? Growth models also try to reproduce in a schematic fashion how that contribution works, although somehow unsuccessfully.

We will start with some basic ideas about models, namely economic models, and an explanation of different types. A model describes a system whether it is the nature, a society or a product of the human mind, though in a limited fashion. An economic model replicates parts of an economic system to simulate its operation, gathering a number of input parameters and the system's output, and linking them through a specific rule. This rule is described as a particular set of relationships, which characterize the model and, hopefully the system. Typically, there are two categories: Static and dynamic models. The former reflects a finite number of closed and necessary relationships between the model's parameters, in the form of mathematical identities. It shows how system parts interact, enlightens the main concepts involved, and clarifies the respective semantics. The latter is an open relationship between system parameters, written in the form of mathematical equations representing a combination of functions, possibly non-linear and time dependent, for which is necessary to find solutions and then to interpret the results appropriately.

A. S. C. Fernandes, *The Contribution of Technology to Added Value*,
DOI: 10.1007/978-1-4471-5001-5_2, © Springer-Verlag London 2013

2.1 Comparative-Static Models

It can be said that accounting systems are also economic models, as they describe the economic activity using units of value, whether they refer to firms, sectors, or national economies. These systems, for both micro and macroeconomics, provide nowadays the structural semantics of this field of knowledge, along with its epistemological component. Unfortunately, this is a main first argument, technology is a concept that is not present within those systems, such that its contribution to production has not been valued so far. Moreover, the accounting systems, being international standards, bring the highest possible objectivity to the economic process, since they ensure the verification of economic facts regardless of place, time, and the observer. Also, they are the only available bridge to interpret how economic value translates into social value. The models based on the (double entry) accounting systems, which are approximately standard all around the world, are static and conservative, what means that the sum of the parts makes the whole, just as energy in Physics or as within Mathematics. These conservative features create the best possible environment to delimit unambiguously the production factors, which somehow relate to technology and to knowledge.

In order to evaluate the evolution of economic systems, thus growth conditions, it is not enough to work with static patterns like accounting identities. It is also required to establish hypotheses between the system parameters along the time. These hypotheses are, after all, conjectures about how the model dynamically relates to its reality context. Examples of conjectures would be the increase in human capital reflecting the increase in fixed capital, or the increase in household consumption being activated by the increase of their income, or even the total factor productivity (TFP) gains representing the earnings that are not measured directly on the contributions of labor and capital inputs. Of course, these hypotheses should be previously verified in the social and economic contexts.

Although the static models provide a static balance, a certain degree of its time dependence can be directly verified. In fact, for a sequence of time periods, it is possible to infer differential models, which, besides relating static parameters representing production, establish relations between their time variations. This generalization of static models allows both the application of a method known as comparative statics and the design of models described by differential identities.

2.2 Dynamic Models

On the other hand, a dynamic model based on a number of time-dependent characteristics, i.e. a time-dependent rule that relates its inputs with the output, can be built. Dynamic economic models are typically referred to as growth models. It will be seen that the neo-classic growth models attempted to include technology in the rule, making an explicit difference with capital, though mixing it with knowledge and technical progress, whatever they mean.

Before resuming to growth models, I will briefly describe what is meant by economic growth and what the fundamental identities in macroeconomics are. The problem of economic growth is a subset of the wider issue of development. The idea of economic growth is basically associated to the increasing accessibility of every individual to goods and services, which will, on the one hand, compensate for the natural fading of his/hers physical and intellectual condition and, on the other hand, provide supplementary capabilities. Therefore, growth relates directly to the increasing yield of an economy and with an appropriate distribution of total income among different identities of a society [1]. Income is at the origin of expenditure, which may have two different forms: Consumption and investment (savings). Consumption benefits the individual in the short run but diverts value from investment. Investment feeds production and may increase output and then income in the medium and long term. As such, the conditions for increasing income are directly related to optimal budgeting between consumption and investment, as seen from the expense view, and with the proper use of resources, from the production view. It is the appropriate balance between consumption and investment that controls the maintenance of a certain growth rate. To find this balance is the main enquire. So, technology, even if only considered as a mix of resources, is surely in the center of the growth problem.

How have the growth ideas and models developed? In Antiquity, the extent of fertile land, the number of subjects and slaves and the size of the armies provided the first measures of wealth. With Mercantilism, wealth was measured at a higher level of abstraction, by the accumulation of currency and precious goods. To achieve that accumulation, the doctrine was to get an export surplus over imports, using customs barriers and promoting exports. Those times emphasized the importance of the monetary capital available, adding this factor to the need of a large labor force.

The models of economic growth appeared with the physiocracy. The physiocrats, named for the supremacy they attributed to the natural order and the Earth elements in the eighteenth century France, reacted against the mercantilist influence on national economic strategies. According to Quesnay's [2] growth scheme, growth was proportional to the cultivated land, from where a free gift was provided, i.e. a value added by the land itself, value that made the difference between the process's starting value, and the ending output value. One can complete this scheme by introducing the capital investment, as Quesnay regarded it, and its influence on the process productivity. The importance of labor and of machinery as capital was clearly acknowledged.

Adam Smith [3] developed, further, the growth scheme of the physiocrats emphasizing the importance of manufacturing to wealth creation, modeling, and describing it in great detail, highlighting many of its peculiarities, naming many of its parameters, and describing its cost structure. Economic growth, according to this author, depended basically on two parameters: Capital accumulation and division of labor, the latter being the responsible for greater productivity. His conception of economic growth, which he did not model schematically, made it clear that growth would depend on the increase of labor productivity being greater than the increase in wages. According to this idea, as explained by Deane [4, point 3], growth could be achieved step-by-step by increases of technical progress: For a certain level of technical progress, economic growth would tend to minimize profits and wages; but, with additional technical progress, thus increasing productivity, profits and wages would rise again toward a new equilibrium. Still, what technical progress was about has not been objectively explained, and the word technology was still absent in the lexicon of those times.

The works published by Harrod [5] and Domar [6], following Keynes' ideas of growth, modeled for the first time the dynamics of economic growth. Even if independently, they followed studies of Lundberg [7] and, according to Imparato [8], of Cassel [9] and Kalecki [10]. Later, they would be continued and developed extensively by the Cambridge School economists, notably Nicholas Kaldor and Joan Violet Robinson, school known as post Keynesian and somewhat neo-Ricardian. Those models are based on three fundamental principles regarding the aggregated parameters of an economy: A balance between supply and demand; production to equal income and income to equal expense; and the central ideas of Keynes that savings equals investment and that this acts as a lever for growth of the expenditure and of the product.

This is the beginning of macroeconomics, the establishment of its fundamental relations and the start of the quest for the determinants of economic growth. The Keynesian school supported the theoretical development until the proposals of Solow [11, 12] and Swan [13], who started a new era that would dominate for 20 or 30 years and has not yet been drained. Later, the endogenous growth theory became popular, as well as the movements that placed knowledge and innovation at the heart of growth.

2.3 Technology and the Solow Model

Solow and Swan presented the first most successful long-run growth model. Swan acknowledged the importance of progress and not just technical progress, drawing attention to the relevance of the administration and other institutions, perhaps with roles as important as technical change. Solow explicitly included technological change [11] or technical change [12] as a basic parameter in his Cobb-Douglas aggregate production function. How does technical change relate with technology? It is not clear, because at the time neither it was well defined. Furthermore, over the course of the next 50 years, it was not objectively established the significance of this technical change factor. To illustrate this, I quote, among many other important authors, Jones [14] who refers to this factor using the following words: *Stock of ideas*; *technology*; *technology variable*; *productivity term*; *stock of knowledge*; *people with ideas*, *labor augmenting technology*, etc. The ambiguity is evident but the current ideas of technology and knowledge are surely contained by it. Solow himself in his 1957 paper explained his technical change parameter as representing any kind of shift in the production function, such as slowdowns, speedups or improvements, or the education of the labor force. It is also important to note that Solow recognized both capital and labor force as explicit factors in his production function, what means that they were conceptually well identified, and that other factors like labor force education and, in general technology were not.

I will explain what the significance of the parameter technical change might be, showing that it is a different thing depending on where it is placed within the Cobb-Douglas's production function. I shall also show how this ambiguity contributed to the difficulty of interpreting some empirical results.

Solow's model builds on two main conditions, both known at the time and already used by other economists, concluding that the economy tends to a steady-state equilibrium when there is no technical change. With modifications in economic conditions, other than capital and labor, the parameter representing technical level will change and new steady-state equilibrium will be reached.

The first condition can be written as a capital accumulation equation, using the idea of comparative statics. It states that the change of the capital value ΔC, from one period of time to the next, equals investment I minus the depreciation of capital during that period of time. Considering differentials to time t, investment I as a part of the product Y, such that $I = s \,.\, Y$, and capital depreciation as $\delta \,.\, C$, the capital accumulation equation is written as in Eq. 2.1, where $\dot{C} = \frac{\Delta C}{\Delta t}$, s is the saving coefficient (dimensionless), δ is the capital depreciation coefficient (dimensionless), and Y is the product of the economy. Typically, Y, C, and ΔC are in units of value, for example €.

$$\dot{C} = s \cdot Y - \delta \cdot C \tag{2.1}$$

This equation reads: In one period of time Δt, the change in capital value ΔC equals the savings part of the economy's product minus the amount of capital depreciated value.

The second condition is a given production function, which describes the relation between the production system inputs with its output Y. This relation corresponds to a conjecture about how the product (output) depends on inputs, i.e. a relationship expressed as a mathematical function in which inputs are independent variables referred here as production factors. Considering as production factors the value of capital C and the number of workers N, the output Y, which is the value of the product, is written as a mathematical function F(C, N). This function, as Solow [12] puts it, represents technological possibilities.

$$Y = F\,(C,N) \tag{2.2}$$

Works of Johan Gustaf Knut Wicksell, Philip Wickstedd, Charles Wiggins Cobb, and Paul Howard Douglas proposed the function F now widely known as the Cobb-Douglas function (2.3).

$$F\,(C,N) = Y = C^{\alpha} \cdot N^{1-\alpha} \tag{2.3}$$

As said above when charactering a model, this function is proposed as the rule that reflects the system operation. What this function says is that, in order to obtain the output Y, capital and labor must operate together in a way that can be represented by a multiplication of their values, each considered not as the whole but only one part of the respective whole: C^{α} as a part of C, and $N^{1-\alpha}$ as a part of N, where α is a dimensionless parameter. This parameter α, typically with a value less than one, balances the contributions to the output of both capital and labor force, meaning that not all capital relates with all workers, only one part of C is worked by one part of N. The rule of this model is this multiplication between the two production factors and the parameter α. It is important to note that a different value of α makes a different function and so a different way the model works. Two models with the same function type but with a different α are not easily comparable, as we will see.

This multiplication means that the output is built by adding $N^{1-\alpha}$ terms, each equal to C^{α}. To make it more clear, suppose $C = 64$ €, $N = 16$ workers and $\alpha = 0.5$. The output would be built by adding four times ($=16^{0.5}$) the value of 8 ($=64^{0.5}$), yielding 32. The units of the result are [$€^{0.5}$.$workers^{0.5}$]. This production function is dimensionally incorrect as the first member has the dimension unit of value [uv], for example €, and thus the second member should have the same dimensions, which it does not. There is, here, a missing parameter A, with appropriate dimensions, which would make the equation correct.

Swan used it in this incorrect way and many authors continue to do so, what represents a serious theoretical matter and a more or less important pragmatic issue as it affects the interpretation of results in an unknown way. Other dimensions have been used for this type of production function, as stated in many economic text books. For example, if dealing with the electricity sector, N may be hours worked by labor, C hours worked by the turbines and Y in units of energy like MWh. For instance, Kaldor [15] proposed steel contents, in tonnes, within the capital assets. It would be wrong again. Y could be tons of cotton, C the value of capital and N the number of workers. It is still wrong. Most authors apparently forget to make this production function dimensionally correct. Hunten [16] makes an excellent short biography of the difficulties found by the empirical literature on TFP and the Solow residual, though without identifying this dimensions error. He quotes Robert Solow in 1987: “We can see the computer age everywhere but in the productivity statistics”. The result of this dimension’s inaccuracy is a story of apparent successes and many inconsistencies within empirical econometric analysis. Next, I will try to clarify this dimensions’ problem, as it deals directly with the measurement of what many authors understand as technical progress and technological change.

Solow [12] starts explaining his theory of the convergence for steady-state equilibrium with a production function not fully defined but equivalent to Eq. 2.3, in the form of Eq. 2.4. Y_N and C_N are product and capital per worker.

$$Y_N = {C_N}^{\alpha} \tag{2.4}$$

The combination of Eqs. 2.1 and 2.4 leads to his famous steady-state equilibrium. Eq. 2.4 is dimensionally incorrect. Still, for comparisons of different solutions, how right or wrong could this be, in view of this problem of the missing parameter A? It is right if that parameter A does not change its value over time when Y, C, and N change. It is wrong if it changes, and the most I can say is that it will be less right if it changes a little and more wrong if it changes a lot.

In Solow's paper section IV, example 3, the following production function is also proposed: $Y = \alpha^2 C + N + 2 \alpha \sqrt{CN}$, where this problem of dimensions is very obvious, as Y and N do not have the same dimensions (the labels were changed to the current notation).

Then Solow introduced another formulation in section VI, and later in Solow [12], by entering the parameter A representing something related to knowledge and technology, which he said that represented technological change. This could be the missing parameter that would put the equations right. However, the ambiguity with which this parameter was introduced did not allow for a clarification, actually it presented extra problems.

The Cobb-Douglas production function was written and used by Solow and many authors placing the technological change parameter A in three different positions: *Augmenting labor*, *augmenting capital*, or *neutral*. The three respective forms are written as follows:

$$Y = C^{\alpha} \cdot (A \cdot N)^{1-\alpha} \tag{2.5}$$

$$Y = (A \cdot C)^{\alpha} \cdot N^{1-\alpha} \tag{2.6}$$

$$Y = A \cdot C^{\alpha} \cdot N^{1-\alpha} \tag{2.7}$$

A dimensional analysis immediately shows that, in each case, A has different dimensions and so must signify different causes, what should have different effects.

For *labor augmenting* (2.5), the dimension of A is units of value per number of workers [uv/nw].

For *capital augmenting* (2.6), the dimension of A is $[uv/nw]^{(1-\alpha)/\alpha}$.

For *neutral* (2.7), the dimension of A is $[uv/nw]^{(1-\alpha)}$.

It must be concluded that this technological change parameter is only exogenous in the *labor augmenting* case (2.5), where its dimensions do not depend on α. Note, as explained above, that exogenous parameters are those that do not depend on the model rule. On the other two cases, A dependence on α means that its economic significance is depending on the model characteristics such that, for different values of α, A means a different thing. Moreover, it should not be referenced with the same symbol. In order to investigate further their meanings, we will use A_1, A_2, and A_3, respectively.

This analysis suggests that the three technological change factors may be a type of labor productivity, measured as units of value per worker [uv/nw] to the power 1, $(1 - \alpha)/\alpha$, or $(1 - \alpha)$, respectively. But they could also be a kind of capital productivity, measured as units of value per worker [uv/nw] to the power 1, $(1 - \alpha)/\alpha$, or $(1 - \alpha)$, respectively.

2.3.1 Technological Change as a Type of Labor Productivity

Let us first consider them as charged labor productivities, meaning labor productivities affected by a charge or a weight. What type of charge is this? And on what is that charge depending? To answer these questions, we will first take the classic and consensual definition of labor productivity LP = Y/N, which unquestionably has dimensions [uv/nw].

In the first case (2.5), $A = A_1$ has the dimensions [uv/nw], the same as LP's. This means that it is a labor productivity (charged) multiplied by a dimensionless factor f_1. In other words, $A_1 = LP.f_1$.

To calculate f_1 we write LP = Y/N and substitute Y by Eqs. 2.5, yielding 2.8. For A_2 and A_3, and f_2 and f_3 factors, the equivalent results are shown in Eqs. 2.9 and 2.10.

$$f_1 = {C_N}^{-\alpha} \cdot {A_1}^{\alpha} \tag{2.8}$$

$$f_2 = {A_2}^{1-\alpha} \cdot {C_N}^{-\alpha} \tag{2.9}$$

$$f_3 = (C_N)^{-\alpha} \tag{2.10}$$

This shows that the factor f is a multifaceted entity with a meaning not easy to understand. Accordingly, A is not really a labor productivity, meaning that it does not relate output value to labor activity in a straightforward fashion.

2.3.2 Technological Change as a Type of Capital Productivity

Let us now inquire the parameters A as charged capital productivities and find out on what are they depending. Again, we will first take the classic and consensual definition of capital productivity CP = Y/C, which unquestionably is dimensionless. We will follow the same method as above.

In the first case (2.5), $A = A_1$ has the dimensions [uv/nw]. This means that it is a capital productivity multiplied by a factor f_1, which dimensions are [uv/nw]. In other words, it would be $A_1 = CP \,.\, f_1$.

To calculate f_1 we write CP = Y/C and substitute Y by Eq. 2.5, yielding $f_1 = {A_1}^{\alpha}.{C_N}^{1-\alpha}$. For A_2 and A_3, and f_2 and f_3 factors, the equivalent results are: $f_2 = {A_2}^{1-\alpha}.{C_N}^{1-\alpha}$; and $f_3 = {A_3}^{1-\alpha}.{C_N}^{1-\alpha}$. Again, this shows that the parameter A is not capital productivity, not even charged capital productivity.

From this brief yet objective analysis the conclusion is that: (1) The parameter A most probably reflects other factors than C and N; (2) however, it is not independent of C and N.

This has important implications on the calculations of the TFP, what will be analyzed in the next section. For now, we may conclude that the neo-classic parameter A, referred to as technical progress, technology, or as everything else contributing to output besides capital and labor, is not independent of C and N and does not inform on the role of technology. One aspect was, however, made clear: Neo-classic and Keynesian authors clearly established a conceptual difference between capital and technology, defining and quantifying capital but leaving technology as an indefinite, and confusing idea.

2.4 Technology and Total Factor Productivity

The objective of computing TFP is to assess the importance and contributions to the output of everything else than labor force N and capital C. That residual part would include workers' knowledge and new ideas, technological assets that could not be considered as capital (disembodied technology), management structure, organizational forms, impacts from markets, financial costs, profits, etc. What neoclassic authors called technology was diffusedly within that lot, though the most significant part. Taking the production function (2.7), the parameter A may play that role. In fact, even if with variations, the literature of growth accounting, productivity growth, or equivalent nominations takes this production function with *neutral* parameter A as the basis for calculating this productivity. Assigning to A the idea of TFP, it reflects the concern of identifying a parameter that relates the output Y with the whole of inputs, and so revealing the influence of everything else besides known and objective inputs, like C and N.

How can TFP be computed? Taking three time series of Y, C, and N for n years, it is possible to construct two time series of Y_N and C_N, and scatter n points, one for each year, on a two-dimensions plane (Y_N,C_N). Note that (2.7) may be written as (2.11). Using the method of ordinary least squares or least squares fitting, it is possible to find the function $Y_N(C_N)$ that estimates and describes how Y_N depends on C_N. Prescribing a power function as the estimator (as the production function suggests), the regression algorithm will return the best fit function, and with it the values of A and of α. The value of A will be the value of TFP.

$$Y_N = A \cdot {C_N}^{\alpha} \quad (2.11)$$

Alternatively, taking logarithms of (2.11), the Eq. 2.12 can be written. In this case, the estimator function would be linear and the regression algorithm will return the values of ln A and α, the former being the intercept, the point where the linear function crosses the ln Y_N axis (ln $C_N = 0$), and the latter being the inverse tangent function of the angle between the linear function and the ln C_N axis.

$$\ln Y_N = \ln A + \alpha \cdot \ln C_N \quad (2.12)$$

This is perfectly equivalent to the first method. In some literature, TFP is said to correspond to ln A instead of A. In one or in the other way, for the time period corresponding to the time series, the values of TFP are computed as described. A conclusion relevant to our analysis is that A and α are a pair of solutions, meaning that their values depend on each other and on the input distributions.

Alternatively, it is possible to force exogenously a specific value of α and compute the linear regression, what would return a different value of ln A. Or, force a value of ln A and have a return of a different value of α. In both cases, there will be higher minimum square errors (MSE), such that the production function would not represent the real distribution equally well.

An example will help to understand this problem of comparing values of A in different periods of time and between different countries. Taking time series for

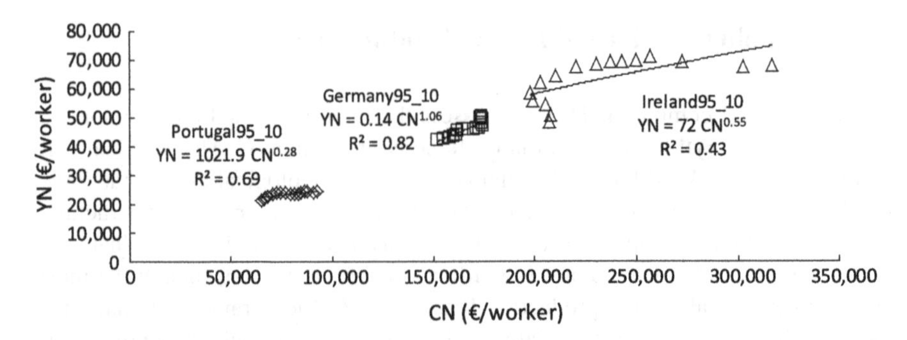

Fig. 2.1 Description of Portugal, Germany, and Ireland economies according to production function (2.11), 1995–2010, constant prices of 2005, data from AMECO

Table 2.1 Values for A and α—Portugal, Germany and Ireland economies according to production function (2.11), 1995–2010, constant prices of 2005, data from AMECO

	$A\ [(€/N)^{(1-\alpha)}]$	α
Portugal	1021.9	0.28
Germany	0.14	1.06
Ireland	72.0	0.55

the Portuguese, German, and Irish total economies, from 1995 to 2010, of Y_N and C_N, where Y is the GVA at constant prices (2005) for total branches,[1] C is the net capital stock at constant prices for total economy,[2] and N is the total labor force, from the European annual macroeconomic database AMECO,[3] the results for the three countries, according to the production function (2.11) are depicted in Fig. 2.1 and listed in Table 2.1.

The values of A for each country in the same period of time, which could be interpreted as the TFP, are extremely different and, as we have seen in Sect. 2.3, their units are different hence not comparable. It is obvious from this example that A is a different thing in each of the countries studied here.

If instead of comparing different economies, along the same period of time, we would compare different periods of time for the same economy, the results would be equally illusive. Taking the same three economies and calculating in the same way, the values of A and α for the periods 1995–1999, 2000–2004, and 2005–2009, the results are shown in Table 2.2. Again, they are very different and not comparable.

It is very clear that these results do not show what it was expected. In no way either it is possible to interpret the parameter A as a productivity that could be comparable between different economies or different periods of time, or interpret it as reflecting anything like technology changes or technical progress. So why is

[1] Definition (ESA 95 [17]): 8.11–8.12, 9.23, 10.27–10.30.

[2] Definition (ESA 95 [17]): 6.02 f.

[3] http://ec.europa.eu/economy_finance/db_indicators/ameco/index_en.htm (consulted March 2012).

Table 2.2 Values for A and α—Portugal, Germany and Ireland economies according to production function (2.11), constant prices of 2005, data from AMECO

	1995–1999		2000–2004		2005–2009	
	A $[(€/N)^{(1-\alpha)}]$	α	A $[(€/N)^{(1-\alpha)}]$	α	A $[(€/N)^{(1-\alpha)}]$	α
Portugal	0.169	1.06	1.564E + 05	−0.17	1,826.214	0.22
Germany	0.004	1.3	3,361.021	0.21	0.000003	1.95
Ireland	1.696E + 20	−2.9	2.51	12.3	0.7	13.2

this methodology being used by many authors and organizations? Possibly for two reasons:

First, because the values of A are not usually calculated, but in its place the yearly differences of A are computed, such that it is not the value of TFP, but the TFP growth rate that is calculated. This shades this problem of the dimensions of A being dependent on the values of α.

Second, because many authors, as Solow [12] himself did, consider the parameter α as input to the system.

We shall start by analyzing the first. In fact, taking derivatives to time of (2.11) and then dividing by Y_N, the equation can be written as in (2.13), where the cap symbol over a parameter means growth rate. The equation reads as follows: The output per worker growth rate equals the input capital per worker growth rate, multiplied by the factor α, plus the TFP growth rate $\hat{a}$. As each member of this equation and the two terms of the second member have dimensions of $[t^{-1}]$, the problem of the dimensions of A dependence on α becomes opaque, though it is still there.

$$\widehat{y_N} = \alpha . \widehat{c_N} + \widehat{a} \tag{2.13}$$

Let us analyze one example. The following TFP growth rate calculation is equivalent to what was presented in Table 2.2, though using periods of 1 year instead of 5 years. Using Portugal data as before and Eq. 2.11, Fig. 2.2 shows the results of the production function for ten periods of one year, from 1995–1996 to 2004–2005. From each production function of each year, values of A and α are extracted and shown in Table 2.3, columns 2 and 3.

The calculations of the TFP growth rate, displayed in Table 2.3, show once again that the yearly TFP growth rates do not seem comparable, and the reason is what was pointed out above: As the dimensions of A depend on α, and α is different from period to period, the values of A in each period represent different things.

Now analysing the second. Solow valued it as the ratio of the return to capital to total return, and for output he used the gross domestic product (GDP). A similar approach is followed today by the Organisation for Economic Co-operation and Development (OECD). This approach makes the problem much smoother in what concerns the large variations of either A or â that we have seen in the cases above, but does not change the nature of the problem.

Solow [11] considered α = 0.35, value that changed every year proportionally to the ratio (labor force)/(labor force employed), from 1909 to 1949. The empirical

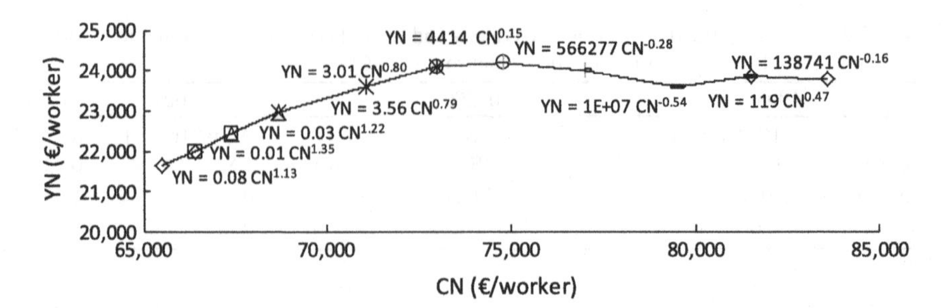

Fig. 2.2 Description of the Portugal economy according to production function (2.11), constant prices of 2005, data from AMECO. Results for ten periods of one year are shown, starting on 1995–1996, on the left, up to 2004–2005, on the right part of the figure. In each period, the production function shown is estimated with only two points, where from the values of A and α are extracted to columns 2 and 3 of Table 2.3

Table 2.3 Second and third columns show values for A and α for Portugal economy according to production function 2.11, constant prices of 2005, data from AMECO. Values are taken from equations on Fig. 2.2. Fourth and fifth columns show â and α, the latter as the average for two consequential periods

	A	α	$\Delta A/A = \hat{a}$	α
1995–1996	0.08	1.13		
1996–1997	0.01	1.35	−0.88	1.24
1997–1998	0.03	1.22	2.00	1.29
1998–1999	3.56	0.79	117.67	1.01
1999–2000	3.01	0.8	−0.15	0.80
2000–2001	4,414	0.15	1465.45	0.48
2001–2002	566,277	−0.28	127.29	−0.07
2002–2003	1.E + 07	−0.54	16.66	−0.41
2003–2004	119	0.47	−1.00	−0.04
2004–2005	138,741	−0.16	1,164.89	0.16

values of α evolved between 0.312 and 0.397. Using an equation equivalent to (2.13) and time series for Y_N and C_N, he could calculate $\widehat{y_N}, \widehat{c_N}$ and thus â as the so-called yearly rate of technological change. Considering α as an input, and knowing that its value for a specific country or sector does not change much along the time, makes the dimensions of A changing less and so making yearly comparisons probably more acceptable. How acceptable? That depends very much on the values of A and α, as their relation is not linear but a power function. It cannot be said, a priori and not easy to assess a posteriori. For example, it will not be possible to compare TFP growth rates between sectors and countries with very different mean values of α.

Moreover, an almost constant α, which geometrically means an almost constant angle between the linear function and the horizontal axis, may have a destroying effect on the regression technique, such that the chosen linear function will no longer represent that specific distribution (even if with only two points).

OECD [18] uses as inputs yearly values of α for evaluating what they refer to as multifactor productivity (MFP)[4] (based on value added). This organization uses an equation, somehow equivalent to (2.12) and representing yearly changes as in (2.13), which can be written as (2.14). The subscripts t and $t-1$ mean year t and year $t-1$.

$$\mathrm{In}\frac{y_t}{y_{t-1}} = \mathrm{In}\frac{A_t}{A_{t-1}} + \alpha \cdot \mathrm{In}\frac{CS_t}{CS_{t-1}} + (1-\alpha)\cdot \mathrm{In}\frac{N_t}{N_{t-1}} \tag{2.14}$$

This equation could present the same problems as before; however, this methodology uses, instead of capital C, what they call capital services CS, which are measured in working hours [19]. Concisely, these services, multiplied by prices that are paid for 1 hour use of each type of service, correspond to a capital use value, or remuneration for capital services, in units of value. The latter, divided by the total costs of inputs, is the value of α; and total costs of inputs are considered to be the sum of the remuneration for labor w . N and the remuneration for capital services. N is the number of hours worked and w is a wage per hour worked (prices and wages are considered in such a way that the final equation is written in constant prices). As such, the basic production function that OECD uses is (2.15), and a dimensions analysis shows that A (what they call MFP based on value added) is measured in [uv/hw] and so becoming independent of α. This methodology overcomes the main problem pointed out before.

$$Y = A \cdot CS^{\alpha} \cdot N^{1-\alpha} \tag{2.15}$$

What now does it measure this MFP? To answer this question we can consider Y, which is the GVA, as the sum of its two most important terms, according to the fundamental accounting identity from the production view (2.16).

$$Y = w \cdot N + m \cdot C \tag{2.16}$$

I call m the management factor [20], which, proportionally to capital C, represents everything else besides the contribution of labor (compensation of employees) to GVA. It covers capital depreciation and revaluations (or consumption of fixed capital), taxes, and other financial costs and operating surplus. So, any ΔA (or the MFP growth rate) accounts for changes in wages, capital service prices, financial costs, and operating surplus. Where is technology or technical change here? It is diluted in all these terms. So, once again, the so-called methods computing MFP do not inform on the influence of technology, whatever that could be.

2.5 Technology and A.C Models

There is a special class of models that are based on production functions of the type shown in (2.17), where the product Y, the model's output, is proportional to capital C, an input to the model and the sole explicit production factor, being the proportionality given by the parameter A.

$$Y = A \cdot C \tag{2.17}$$

[4] Definition in http://stats.oecd.org/glossary/detail.asp?ID=1698 (Accessed March 2012).

The parameter A has had interpretations as in the neo-classic model. Thus, A reflects the knowledge and technology within the system, one or the other depending on the author, and C represents capital with the variant that it can include human capital as well. As pointed out before, these ambiguities continue to challenge an objective interpretation of results.

This type of models is characterized by increasing returns to scale, for example, if C and A double, Y more than doubles, which was not the case in the Solow model, where the economy had decreasing returns to scale in its dependence on capital. The interpretation of A is indistinct but its dimensions are clear. If C is capital, measured in units of value [uv], as well as the product Y, A comes dimensionless. Its meaning is a clear capital productivity, $A = Y/C$, everything in the system that contributes to produce Y out of the use of capital C, in other words, labor and their skills, technology (whatever it means), technical progress in organizational structures, and other contributors to output like financial costs and profits.

This model is handled with the consensual capital accumulation Eq. 2.1, which says that the capital increase in each period of time equals the investment I, or savings $S = I = s . Y$, less depreciation of the existing capital $\delta . C$. Dividing (2.1) by C, considering (2.17) and writing the corresponding expression in terms of growth rates originates (2.18).

$$\widehat{y} = s \cdot A - (\delta - \widehat{a}) \quad (2.18)$$

The product's growth rate $\widehat{y}$ comes proportional to the saving coefficient s. An economy with constant technical progress A, what can be represented by $\widehat{a} = 0$, and a constant depreciation coefficient $\delta = \text{const.}$, grows only according to savings increases. Similar conclusion had been taken from Harrod/Domar model's results, where the equivalent of A was named by Domar [6] as *potential social average investment productivity*, which, in turn, was a concept close to the famous Keynesian multiplier [21], Chaps. 3, 8, and 10).

Let us see how A behaves in such a model. Writing (2.17) per number of workers does not change the meaning and the values of A, and we can write (2.19):

$$Y_N = A \cdot C_N \quad (2.19)$$

Taking the same economies along the same period of time as in sections above, the results for A are depicted in Fig. 2.3 and Table 2.4 (linear regressions were forced to an intercept equal to zero).

The results show average values for the capital productivity about the same for the three economies, during this period of time. As there are no differences in dimensions of A, these values can be compared and the comparison is significant.

In the same way, the capital productivity growth rate can be computed using periods of 1 year and evaluating $\widehat{a} = \Delta A/A$. Using Portugal data as before and Eq. 2.19, Fig. 2.4 shows the results of the linear production function for ten periods of one year, from 1995–1996 to 2004–2005 (linear regression were forced to a zero intercept). From the production function of each period, the value of A is extracted and shown in Table 2.5.

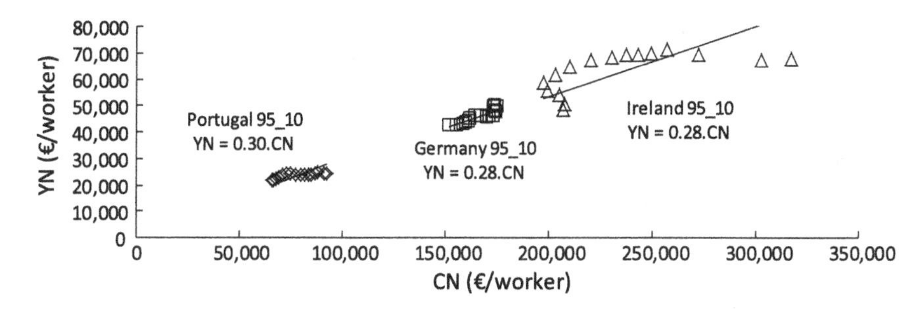

Fig. 2.3 Description of Portugal, Germany, and Ireland economies according to production function (2.19), 1995–2010, constant prices of 2005, data from AMECO

Table 2.4 Values for A—Portugal, Germany and Ireland economies according to production function (2.19), 1995–2010, constant prices of 2005, data from AMECO

	A
Portugal	0.30
Germany	0.28
Ireland	0.28

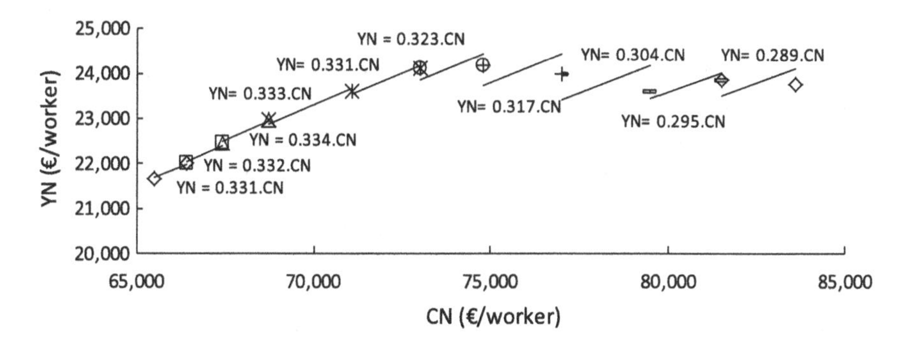

Fig. 2.4 Description of the Portugal economy according to production function (2.19), constant prices of 2005, data from AMECO. Results for ten periods of one year are shown, starting on 1995–1996, on the left, up to 2004–2005, on the right part of the figure. In each period, the linear production function shown is estimated with only two points, from where the values of A are extracted to column 2 of Table 2.5

These numbers for the capital productivity illustrate how the A.C dynamic model may describe an economy, rigorously though in a very limited fashion. The parameter A, used in these models, has a clear significance, is dimensionless, and reveals correctly and unambiguously how capital relates to output. However, from the point of view we are pursuing, which is the influence of technology on growth, it is still very ambiguous.

Table 2.5 Second and third columns show values for A and $\hat{a} = \Delta A/A$ for Portugal economy according to production function (2.19), constant prices of 2005, data from AMECO. Values are taken from equations on Fig. 2.4

	A	ΔA/A
1995–1996	0.331	
1996–1997	0.332	0.003
1997–1998	0.334	0.006
1998–1999	0.333	−0.003
1999–2000	0.331	−0.01
2000–2001	0.323	−0.02
2001–2002	0.317	−0.02
2002–2003	0.304	−0.04
2003–2004	0.295	−0.03
2004–2005	0.289	−0.02

2.6 Endogenous Technology Models

In previous neo-classic models, technology, technical progress, and knowledge are accounted for within one or two parameters (A and α for the Solow model and A for the A.C model). In the Solow model, with its Cobb-Douglas production function, α is an input as well as labor and capital, and A is said to be exogenous, meaning that its value cannot be computed only from inputs. The same applies to A.C models. As most authors considered A as somehow representing technology, those are known as exogenous technology models.

Romer [22] introduced a new model in which the variable that reflects knowledge and technology is a function of some of the model′s inputs. This and other similar models are known by endogenous growth or endogenous technology models. He devised a model with three activity sectors: First, a sector where ideas for new goods are produced, characterized by an R&D activity, with a population of researchers N_A; second, a sector that produces capital equipment C that results from the ideas developed in sector 1, and that does not have labor force; third, a sector that produces final consumption goods, with a labor force N_C and using the capital produced in sector 2. All goods from the first sector are used in the second and all goods from the second are used in the third. This last sector sells its products to the market. The output product of this economy is produced exclusively by the third sector.

In the final consumption goods sector, the production function used is neoclassical, with constant returns to scale. If C is the capital used by this last sector and Y is the economy output, the model's production function is (2.20), in its simplest form:

$$Y = C^{\alpha} \cdot (A \cdot N_C)^{1-\alpha} \tag{2.20}$$

A is now an index that reflects the stock of knowledge or ideas, ideas that are generated by N_A researchers within sector 1. Researchers discover new ideas in every period of time, such that A increases with time, with the number of researchers N_A and with H, an equally important parameter reflecting human capital, such as education years of every worker. On the other hand, each idea corresponds to a unit of capital x, starting with A ideas and growing every year. Thus, capital C becomes proportional to A . x and growing as $\dot{A}$.

Overall, the model works as a one sector neo-classical model with technological change but with an endogenous explanation of the source of technological change. This is triggered by R&D activity, a non-rival good, which immediate costs are well overcome by overall social benefits.

As in the neo-classic models, the endogenous technology models take that residual idea of everything else besides capital and labor, and name it, definitely, as technology. This technological portion is expressed as a combination of human capital, taken as a measure of workers' knowledge, with labor activity, measured as a number of workers, and also with capital formation, representing embodied knowledge. So, the idea of technology is still used overlapping the ideas of knowledge and capital but there is already a timely sequence showing how knowledge originates first more productive work, and later how this generates new forms of capital.

2.7 A New Linear Model

As shown above, along the twentieth century, the idea of technology became central to the process of value adding; consequently, it was progressively considered in production functions describing production processes. Labor and capital are two inputs that have been quantified explicitly, weighted differently in the production function and represented as one multiplied by the other. Technology, because it could not be defined objectively, has not been an explicit input. In fact, how could it be assigned a quantity or a value to a still ambiguous idea? The idea of technology was within a mix of everything else besides labor and capital contributing to output. That mix is said to include workers knowledge and skills, organization structure, and technical solutions. In reality, it is more than that because financial costs and operational profits also contribute positively to output (GVA).

Suppose, as a working hypothesis, that we could define technology such that its concept would point to an autonomous entity, and thus allowing a quantification and valuation of what I will call technological assets TA. Also, as a corollary, that we could objectively discriminate technological assets from capital assets CA, and finally that we would understand the differences and the borders between human knowledge, work, and technology. If we could do that, we should be able to consider, independently, three inputs to a production process: Labor, technology, and capital. In Chap. 3, it will be described how these new operational concepts are constructed such that this working hypothesis is justified. For now, I propose to follow a sequence of economic events, increasingly more complete, sequences that describe production systems and their models, which will enlighten the process of value adding when using human knowledge, work, technology, and capital contributions as inputs. This will introduce a step-by-step new production model with a linear production function, which I believe is simple, intuitive, and thorough.

2.7.1 A Simple Production Model

The simplest economic model involves only one primitive family and the surrounding Nature. The family has their own knowledge, which, in absence of any tools, comprehends only the ways and strategies to remain alive and to prosper. Naturally, families weaken without consumption. That fading away is both physical degradation and information decay. In the limit, if the family would not eat and internalize information would degrade their organic constituents, decompose, and disperse to nature. On the other hand, if the family works in accordance with her knowledge, will obtain from Nature what is needed to compensate for the natural degradation, thus keeping its ability to live and thrive.

I will consider stocks and flows to explain how models work. Moreover, as I will be talking about entities with apparently different natures, it will be used a single quality for all, which is economic value. As such, there will be stocks and flows of value.

In this first model, there is a flow of knowledge value from the family to Nature, carried by action and work of the family members. The result of that work is whatever is harvested from Nature, and the product of that process, the goods produced, goes back to be consumed by the family. This is an economic cycle. If the work value equals the goods value, the family will be in equilibrium. If it is larger, the family will lower its capacity to survive; and if it is smaller the family will have the chance to grow. In this model, the only entity that can accumulate value is the family and only in the form of more knowledge. There is value accumulation when the flow in the family's direction exceeds the one in the direction of Nature. This is outlined in Fig. 2.5.

If the stock of knowledge grows, it is foreseeable that, in the same period of time and in comparable circumstances, the goods produced increase beyond the basic consumption daily needs, what will lead to an increasing stock of goods. If work consists of hunting and gathering food, the values are equal when the amount of work produces the goods needed to keep the community at the same level of knowledge, i.e. feeding and dressing enough to compensate for the natural wear and tear.

This is the reference model. However, we can now consider that there may be a surplus of food and furs. Any surplus of food and clothing is a surplus of value created, which, not being consumed, can be kept allowing future exchanges for other products of other families' surpluses. Thus, under these new circumstances, there is accumulation of value in more forms than just human knowledge. The quantity and the value of knowledge existing in the human mind cannot be directly counted, but human communities realize that who uses more efficient techniques

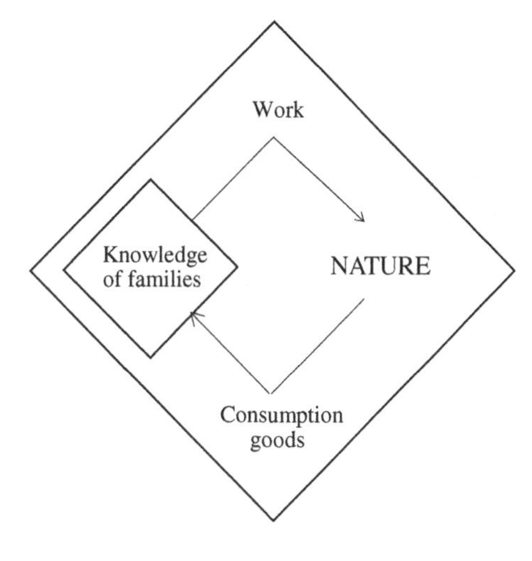

Fig. 2.5 Economic model for families and Nature. *Arrows* indicate value flow directions

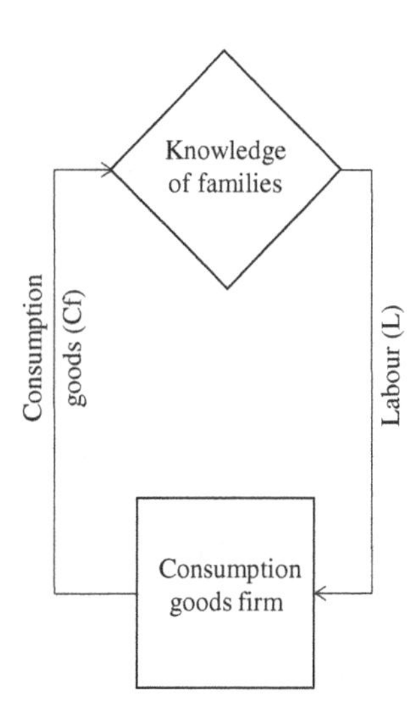

Fig. 2.6 Economic model for families and consumption goods firms. Nature remains as a background. *Arrows* indicate value flow directions

contributes to a higher value work, precisely because, in the same period of time and in comparable circumstances, can generate extra surpluses.

A similar model considers families and Nature and a new entity, which I call a consumer goods firm. This firm is a group of people with an explicit social objective: To produce specific consumer goods. In a primitive economy, such a firm can be a group of hunters, shepherds, or farmers. For simplicity, several of those groups, or firms, are here reduced to a single firm, and it is assumed that they do not have any technical tools besides their own knowledge. The production process remains as it was explained above and the model is described in Fig. 2.6.

Families contribute to firms with the work value L and the firms return to families the value of consumer goods C_f. In static equilibrium, the value flows are

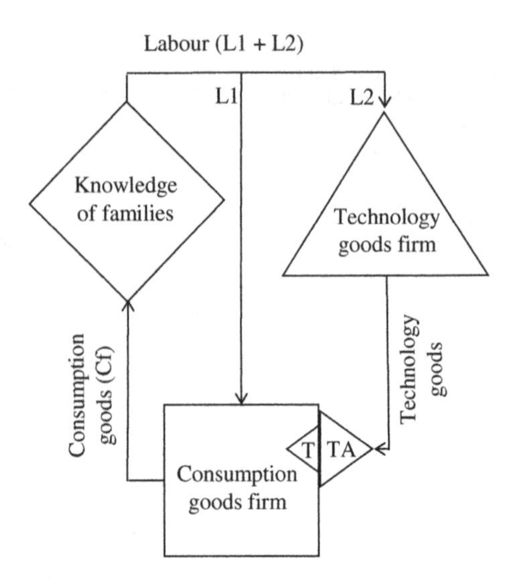

Fig. 2.7 Economic model describing value flows among families, firms producing consumption goods, and firms producing technological goods. TA is the technology goods value that is now available for consumer goods firms. T refers to the part of TA that is used by the consumer goods firm in one economic cycle

equal and the added value can be measured in either flow. The firm, by hypothesis, cannot accumulate value. Once again, the only entities that can do so are the individual families. Nature gives away, at no cost (no value flows), all the raw materials. The value added Y equals the consumption C_f and equals the work value L.

$$Y = C_f = L \tag{2.21}$$

2.7.2 A More Complete Model

If the goods produced are not fully consumed there will be savings. They can be used to "finance" extra work besides hunting and gathering, for example, tools manufacturing. This situation would lead to a new entity, a tools producing firm. Those new groups could be dedicated, for example, to the manufacture of tools for treating animal skins or the preparation of spears for hunting. These tools are conceived out of families' knowledge and work experience and embodied in material forms. We will call these groups as technology firms, because they develop and produce technological assets with value TA.

A new model, described in Fig. 2.7, shows how value flows among these three entities: Families, consumption goods firms, and technology goods firms. It is assumed that both families and consumer goods firms can accumulate value, in the form of knowledge and savings and as technology assets, respectively. Again, furnished by Nature, raw materials are a *don gratuit.*[5]

The families work is now divided by consumer goods and technology goods firms. The labor value flows from families to these firms ($L = L1 + L2$). The

[5] As it was said in 1758 by Quesnay [2] in his *Tableau Economique.*

technology firms deliver their technological assets value TA to consumption goods firms. In the technology firms, the labor value input is L2, which equals the output value TA, hence L2 = TA.

Consumption goods firms deliver their output to families, which value is C_f. However, unlike in the previous case, the consumer goods value C_f is greater than the labor value L1. C_f equals to the sum of labor value L1 plus a value T, i.e. $C_f = L1 + T$. The value T may be less or equal to TA and therefore less or equal to L2. The remaining part of technology assets value (TA-T), if exists, is accumulated in the consumption goods firm.

The value added by this economy Y is the value of the goods produced $Y = C_f + TA$. However, as TA = L2; $C_f = L1 + T$; and L = L1 + L2; therefore, the value added by this economy is also the sum of the value added by work L and the value added by the use of technology T.

$$Y = C_f + TA = L + T \tag{2.22}$$

The families' balance sheet is in equilibrium if the labor value (L1 + L2) equals to consumer value C_f, what will happen when T = TA, in other words, when the technological goods value TA is 100 % used by the consumption goods firm. If not, the families accumulate value in the form of technological assets. In fact, they will have a positive balance sheet when TA > T, hence we could say that the economy grows. The accumulated value would be TA-T, which is stocked in the firm, firm that is the property of families.

Besides the accumulation possibility in the firm's assets, there may be consumption goods accumulation of in the households. In that case, there will be the chance to dispose of extra work for other goals. The families already have one part of the group hunting and gathering, another part manufacturing tools and other technological forms. They may want to form a third group to arrange for new and better living and manufacturing conditions. This could be for example preparing land, building barns, or arranging for barter. I will refer to this third type of activity as producing capital goods.[6] This new situation is described in Fig. 2.8.

Capital goods companies have at their disposal the TA3 technology assets value and the CA3 *capital* assets value, of which they use for their production the parts T3 and C3. With their labor L3 and with the parts T3 and C3 they produce capital goods with value CA = CA1 + CA2 + CA3. Technology goods companies have at their disposal the technology assets value TA2 and the capital assets value CA2, of which they use the parts T2 and C2 for their production. With their labor L2 and the parts T2 and C2, they produce technology goods of value TA = TA1 + TA2 + TA3. Consumer goods companies have at their disposal the technology assets value TA1 and the capital assets value CA1, of which they use

[6] This new idea of capital is somehow different from the current idea of capital, as it differentiates from technology. As such, the current idea of capital C includes the new ideas of technology (assets) TA and the capital assets CA: such that C = TA + CA. The new (capital) is written underlined.

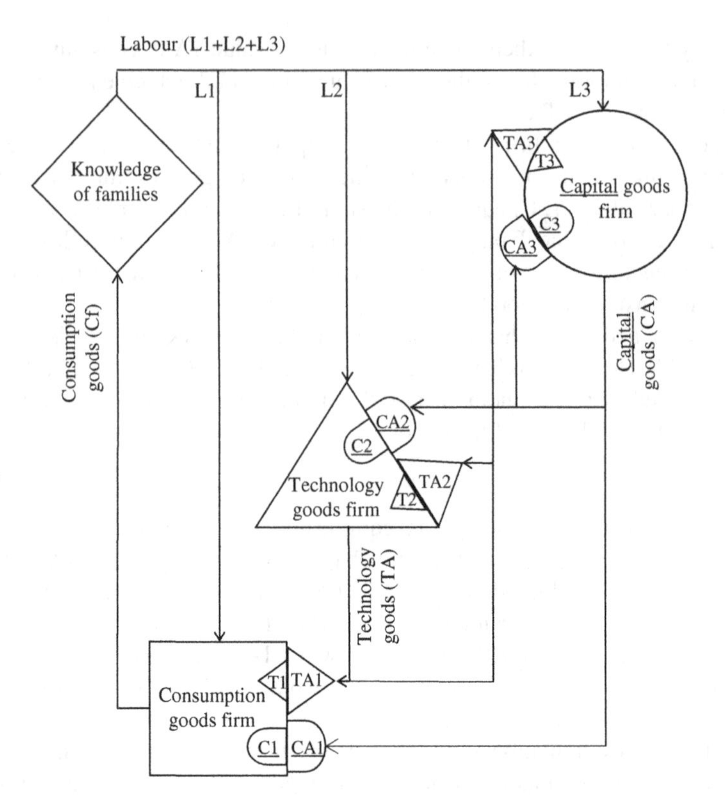

Fig. 2.8 Economic model describing value flows among families, firms producing consumption goods, firms producing technological goods, and firms producing capital goods. Technology goods TA are now available for consumer goods firms, for technology goods firms and for capital goods firms. Capital goods CA are available for consumer goods firms, for technology goods firms and for capital goods firms. T refers to the part of TA that is used by the firm in one economic cycle. C refers to the part of CA that is used by the firm in one economic cycle

for their production the parts T1 and C1. With their labor L1 and the parts T1 and C1, they produce consumer goods of value $C_f = L1 + T1 + \underline{C1}$.

This model, even if explained with a primitive economy that was gradually completed, reflects any modern economy. The value added Y in a certain period of time is equal to the sum of the value of consumer goods C_f with the value of the technology goods TA and with the value of capital goods CA, produced in the same period of time.

$$Y = C_f + TA + \underline{CA} \tag{2.23}$$

As $TA = L2 + T2 + \underline{C2}$; $\underline{CA} = L3 + T3 + \underline{C3}$; and $C_f = L1 + T1 + \underline{C1}$, whereas L groups the values of L1, L2, and L3; T groups the technology uses values T1, T2, and T3; and C groups the values of the uses of capital assets C1, C2, and C3.

Then it comes finally:

$$Y = L + T + \underline{C} \tag{2.24}$$

Based on the working hypothesis considered at the beginning of Sect. 2.6, it was shown that the value added in an economy can be calculated as the sum of the value contributions of the uses of three basic production factors: knowledge, technology (assets) TA, and capital (assets) CA. The three terms of the value added Y are respectively labor L, as the expression of knowledge value, the use value of technology T and the use value of capital $\underline{C}$. Expression (2.24) shows a linear production function as the sum of three components, assuming that it is possible to objectively differentiate among knowledge, technology, and capital.

How to differentiate the three concepts and how to objectively quantify their independent contributions to value added is the goal of the next chapter.

References

1. Castells M, Ince M (2003) Conversas com Manuel Castells. Porto: Campo das Letras Editores (Portuguese translation from Conversations with Martin Ince). Blackwell Publishing Ltd, Oxford
2. Quesnay F (1985) Quadro Económico. Lisboa: Fundação Calouste Gulbenkian, 3ª edição. Published for the 1st time in 1758
3. Smith A (1956) Inquiry into the nature and causes of the wealth of nations. Collier and Son Corporation, New York (original publications on 1776)
4. Deane P (1978) The evolution of economic ideas. Cambridge University Press, Cambridge
5. Harrod RF (1939) An essay in dynamic theory. Econ J 49(193):14–33
6. Domar ED (1946) Capital expansion, rate of growth, and employment. Econometrica 14(2):137–147
7. Lundberg E (1955) Studies in the Theory of Economic Expansion. Kelley and Millman, New York (Chapter 1 contains part of his reference work of 1937)
8. Niehans J (1990) A history of economic theory—classical contributions 1720–1980. The Johns Hopkins University Press, Baltimore
9. Cassel G (1921) Theoretische Sozialokonomie. Winter, Leipzig (cited in Niehans 1990)
10. Kalecki M (1935) A macrodynamic theory of business cycles. Econometrica 3(3): 327–344 (cited in Niehans 1990)
11. Solow RM (1956) A contribution to the theory of economic growth. Quart J Econ 70(1):65–94
12. Solow RM (1957) Technical change and the aggregate production function. Rev Economics Statistics XXXIV 3(99):249–283. (In Economic growth in the long run: A history of empirical evidence—vol II The international library of critical writings in economics 76, Bart van Ark (ed) Elgar Reference Publishing, Cheltenham, pp 32–40)
13. Swan TW (1956) Economic growth and capital accumulation. Econ Record 32: 334–361. Also in Economic Growth, The Economic Record 78(243), Dec 2002 (375–380)
14. Jones C (1998) Introduction to economic growth. M. M. Norton and Company, New York
15. Kaldor N (1957) A model of economic growth. Econ J 67(268):591–624
16. Hulten C R (2001) Total factor productivity. A short biography. In: Hulten CR, Dean R, Harper MJ (eds) New developments in productivity analysis. University of Chicago Press, Chicago
17. ESA95 European system of accounts, Eurostat, ECSC-EC-EAEC, Brussels, Luxemburg, 1996

18. OECD (2001) Measuring productivity, measurement of aggregate and industry-level productivity growth - OECD nanual, OECD Publications, France. http://www.oecd.org/dataoecd/59/29/2352458.pdf. Accessed Mar 2012
19. Schreyer P, Bignon P, Dupont J (2003) OECD capital services estimates: methodology and a first set of results. OECD statistics working papers 2003–2006 OECD Publishing. http://dx.doi.org/10.1787/658687860232. Accessed Sept 2012
20. Serzedelo L, Fernandes ASC (2011) Invisible technology—organisational factor. In: Proceedings of the IEEE Eurocon 2011 international conference 27–29 April Lisbon Portugal. IEEE, USA
21. Keynes JM(1983) General theory of employment, interest and money. The collected writings of John Maynard Keynes VII (1st ed. 1936). MacMillan, London
22. Romer PM (1990) Endogenous technological change. J Political Economy 98(5):S71–102

Chapter 3
A Model to Measure Technology

For the proposed goals of quantifying the independent contributions of knowledge, technology, and capital to value added, the methodological approach to be followed can be synthetically explained in three steps. First, a production model and its production function are conceived, such that the only inputs to be considered originate from knowledge, technology, and capital. This implies that the three inputs are independent parameters, and so evaluated in an objective fashion such that their respective quantities, their values, may be added up arithmetically as in any conservative system. Second, to make the first step possible, the current concepts of knowledge, technology, and capital will be reconstructed as independent operational concepts. Third, the new production function will be apposed to the basic standard accounting identity that evaluates the gross value added (GVA) in a production process. This comparison, through an algorithm to be explained, will allow identifying each term of that basic accounting identity with each of the three terms of the new production function.

The first step was already taken in the last section of the previous chapter, as a working hypothesis. Briefly, it is assumed that it is possible to calculate the GVA of an economy by adding up three and only three contributions originated, during the production process, by the use of knowledge, the use of technology, and the use of capital: $GVA = L + T + \underline{C}$.

This assumption was extensively justified. Now, as the second step, it will be necessary to analyze and rebuild the three concepts: knowledge, technology, and capital. These concepts materialize in what are typically named as people and assets. Thus, it is considered that an economy has as resources only these three types of assets: People's knowledge, technological assets (TA), and capital assets (CA). During production, the value contributions from the use of the three assets will be labor L as a direct expression of human knowledge, T the use value of technological assets (TA), and $\underline{C}$ the use value of capital assets (CA).

A. S. C. Fernandes, *The Contribution of Technology to Added Value*,
DOI: 10.1007/978-1-4471-5001-5_3, © Springer-Verlag London 2013

3.1 Operational Concepts and Methodology

A concept, or a conceptual notion, contains the semantic contents of a notion, whereas the operational notion, also a scientific notion, allows the use of operations and procedures, and most importantly permits measurements [3, 8]. No conceptual model is of any use if it cannot be tested against data [9]; for this, a metric will be needed, which includes a criterion and a scale. However, not being the same, it is important to be very clear to establish the epistemic correlation between the conceptual notion and the operational one [5]. It will be considered in this study that an operational concept is a part of the respective current concept and that it can be objectively used in a model or in a procedure. As such, the operational concept can be represented as a parameter that is objectively and autonomously identifiable. In this way, it may be measured using a metric independent of the observer, the place, or the time.

The construction of the operational concepts for technology, knowledge, and capital is briefly described in this section. It is assured that each operational concept represents an independent entity, without interconnections, which associated value may be linearly added, and thus putting aside any super additive problems. The reconstruction is done through an epistemological analysis of each concept followed by a new synthesis, which obeys to predefined criteria. Listing the elements that characterize each of the three current concepts and understanding which the fundamental ones are should allow an essential distinctiveness between them, and consequently the acknowledgement of their different roles in production.

A schematic relation among the three concepts is depicted in Fig. 3.1 [7]. This figure shows Nature as the background, without which the analysis does not make sense, along which endless cycles are described as a simple dynamic model: Human knowledge, along the time, develops technology, and part of which is used to develop capital. As technology and capital assets are built with knowledge and

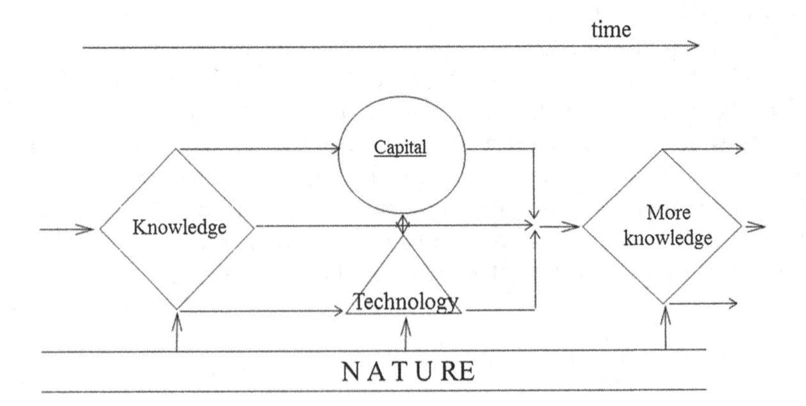

Fig. 3.1 Dynamic links between knowledge, technology, and capital [7], reproduced with permission from Elsevier [reprinted from publication Technological Forecasting and Social Change 79]

work, new goods are produced and consumed, resulting in more and possibly new knowledge.

Annex describes an epistemological analysis of the technology concept, interpreting its current attributes and extracting from them its fundamental elements, such that an operational concept may be reconstructed according to specified criteria. Comparable analyses are performed for the concepts of knowledge and capital which allows establishing the borders, operationally, between knowledge and technology and between technology and capital.

3.2 Technology and Knowledge

The foremost conclusion of the analysis described in annex is that technology is a result of a process initiated by human knowledge, combining action and work with natural resources. As such, it is already a produced good. Also, it was concluded that technology has a well-defined function in a production process, facilitating and multiplying the value of work. A technological form i.e. technology has a material form, which is objective, identifiable, and quantifiable. It represents human's knowledge embodied in material forms, though in a static form, such that it should no longer be referred to as human's knowledge. Explaining the operational concept of technology in this way implies that the operational concept of knowledge is reduced to human's knowledge, meaning the knowledge that exists in the human's mind.

Individual knowledge is a dynamic system of data and information resident in the human's mind, whose dynamics have two important and mutually dependent engines. The first is the permanent reorganization of data and information within the mind, consciously and unconsciously, what may originate new information and new knowledge. The second is the management of bidirectional data flows, between the mind and its exterior—the own body and surrounding nature. Therefore, human knowledge is an individual quality, very dynamic, and unquantifiable. On the other hand, human knowledge is expressed as language, intentional action, and work. The latter is the only aspect that may be quantified. Work will be considered as the quantifiable manifestation of knowledge, and the value that society attributes to labor will thus represent the value added by the use of human's knowledge in a production process.

Consequently, discriminating between technology and knowledge becomes an easier task: (1) Human's knowledge is dynamic and exists within the human's mind; (2) Technology is a produced good where human's knowledge is embedded but in a static form. Therefore, an organized thought, a new idea, a theory, a schematic image, a set of principles, a formulation of a question, a decision sequence, any image or other forms of memorised data, a melody, and so forth are knowledge as far as they exist in the human's mind, because they represent data and information constructed and resident there. On the other hand, a message written on a paper, a sketch, a hand written or CD-recorded speech, a built object, a programmed decision algorithm, a technological rule, as defined by Aken [1], a methodological written sequence and so on, up to tools, instruments, machines, and large systems are technological forms, because they are exterior and already independent of the minds that created them; they are objective forms in a material state. They are available directly by society and are potentially understood and used by many people.

3.3 Capital and Technology

For the operational concept of capital the explanation proposed is the following: Capital forms are objective and identifiable material goods with the same attributes as technology, but with a much larger flexibility in the role they play in a production process. Now, to differentiate technology from capital, it is important first to remember that capital and technology attributes mostly coincide. They are both produced goods not to be consumed, with specific functions in the production process. They have static knowledge content and often a recognizable shape. Only one attribute may contribute to discriminate between them: Capital has a more flexible applicability than technology. Capital's flexible applicability means that it is easily transformable (liquidity). On the contrary, a technology form has a specific function and cannot be used for anything else. Therefore, the boundary between technology and capital is traced with the following criterion: That forms that is versatile in its functionality and is easily transformed is a form of capital. If its function is well defined and cannot be easily transformed, it is a technological form. As such, examples of technological forms are patents, blueprints, reports, tools, instruments, machines, systems and all sorts of equipment, whereas capital forms are typically buildings, money, and credit. Also, land should be considered as a basic form of capital, as far as it is involved in a production process. This border is not absolute, and sometimes the border line is difficult to draw. The algorithm to be proposed to quantify the used values of technology and capital take this difficulty into account.

3.4 The Model's Algorithm

A production process, as a value adding process, is schematically described in Fig. 3.2 where the arrows mean positive value flows. The production process has four input value flows: T, $\underline{C}$, L, and intermediate products value (IPV). They originate in the firm's assets value TA + $\underline{CA}$. The output value is the final product value, FPV = GVA + IPV, which equals the sales' value thus returning to the firm's assets.

The GVA in a firm can be computed as the sum of the value added by the use of the three basic resources: The values added by the direct use of knowledge

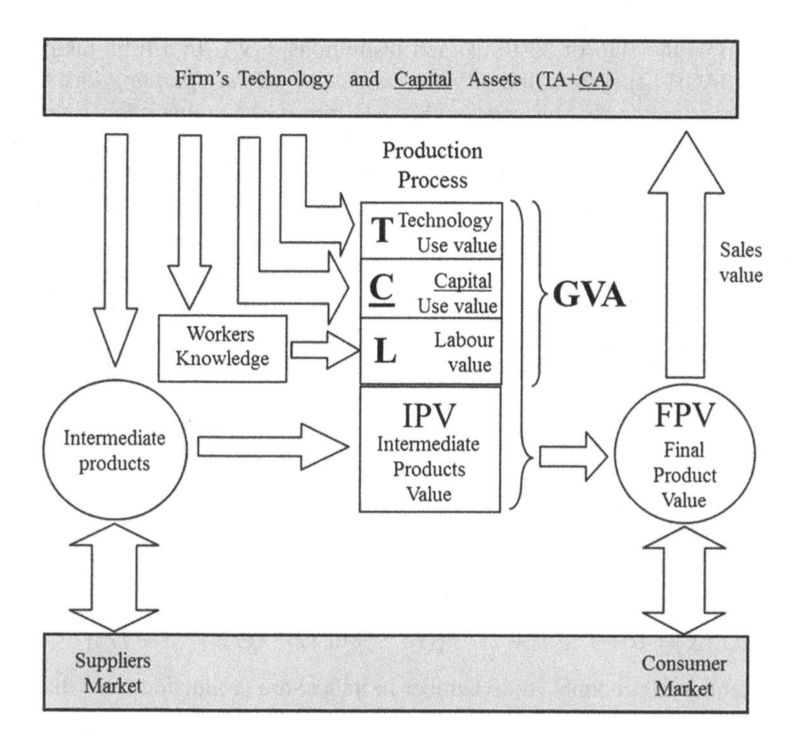

Fig. 3.2 The production process as a value-adding process [7], reproduced with permission from Elsevier [reprinted from publication Technological Forecasting and Social Change 79])

(labour) L, by the use of technology T and by the use of capital $\underline{C}$. As such, the identity (3.1) shows the sum of the three value contributions:

$$\mathrm{GVA} = \mathrm{L} + \mathrm{T} + \underline{\mathrm{C}} \tag{3.1}$$

The GVA is measured in units of value, like euro or dollar; hence, L, T, and $\underline{C}$ are measured in the same units of value. Dividing both members of (3.1) by GVA, the identity (3.2) may be written, where the unity is in the left member of the equation and the right member contains the sum of the three indexes indicating the relative parts of GVA originating from the use of human knowledge, KI = L/GVA (Knowledge Index), from the use of technology, TI = T/GVA (Technology Index), and from the use of capital, $\underline{CI} = \underline{C}/\mathrm{GVA}$ (Capital Index). The indexes are dimensionless and will be less than one, except in uncommon situations.

$$1 = \mathrm{KI} + \mathrm{TI} + \underline{\mathrm{CI}} \tag{3.2}$$

A production process produces goods that society acknowledges as having a certain value. Along that production process, there is value that is added to the IPV: It is the GVA. From the production approach, GVA was defined in Europe by ESA 1995 [6] and equals, in a simplified closed society without state, the national output value minus the national intermediate consumption value. The GVA in an economy is the sum of all firms' and institutions' GVA. In a firm, taking the European BACH [2] system, the GVA is computed by *Total operating income* (*S*) minus *Cost of materials and consumables* (*5*) minus *Other operating taxes and charges* (*8*). However, instead of calculating GVA by a subtraction, it may be computed by summing up the terms that contribute directly: *Staff costs* (*6*), plus *Value adjustments on non-financial assets* (*7*), plus *Taxes on profit* (*Y*), plus *Profit or loss for the financial year* (*21*), plus *Value adjustments on financial assets* (*12*), plus *Interests and similar charges* (13), plus *Extraordinary charges* (*17*), minus *Financial income* (*9–11*), and minus *Extraordinary income* (*16*). The account *Value adjustments on non-financial assets* (*7*) is divided into *7a* (*Depreciation on intangible and tangible assets*) plus *7c* (*Other value adjustments*). Basically, GVA is the sum of four terms: Costs of labor, plus the depreciation of tangible and intangible assets and provisions, plus taxes on profits, plus *profits on the ordinary activity* (*POA*). This last term is the profits on the financial year minus the part of profits originated in extraordinary and financial activity. As a sum of accounts, it is shown in the following identities:

$$\begin{aligned} \text{GVA} &= 1 + 2 + 3 + 4 - 5 - 8 \quad \text{or} \\ \text{GVA} &= 6 + 7 + Y + POA \quad \text{or} \\ \text{GVA} &= 6 + 7 + Y + 21 - [(16 + 9 \text{ to } 11) - (12 + 13 + 17)] \end{aligned} \tag{3.3}$$

The mentioned accounts (code number in italics) are production costs that represent streams of value that flow from the firm's assets to become incorporated in the final product. To distinguish whether each account relates to the use of knowledge L, the use of technology T or the use of capital C, its nature and origin has to be closely analyzed.

The value of the use of knowledge is assessed through the value of work, which is the main objective product of human's knowledge—*Staff costs* (6). No other account contributes to the term L of identity (3.1). The other GVA accounts originate either in technology or capital assets, TA or CA, or in both, thus contributing to either T or C, or to both, as shown in Fig. 3.2.

It was pointed out that there may be an uncertain boundary between technology and capital assets, and thus on their respective use values. The border line, as it was defined, depends on the resource's applicability, being more on the technology side if the resource has a narrower applicability, and more on the capital side when the resource has a wider applicability. Moreover, that difference will depend on the particular firm's production process; for example, the technological content of one firm buildings' depreciation may be different from the one of another firm, which has a different technological index TI. To assess this particular effect the TI will be used and two new parameters will be introduced: The technology coefficient k_a, and the GVA coefficient k_p.

The technology coefficient k_a indicates how much depreciation value from TA is used relative to depreciation of total fixed assets. It is computed as the ratio of the following terms: The first is the depreciation values of plant and machinery, industrial property plus R&D costs; the second is total depreciation value of tangible and intangible fixed assets. The values of k_a are easily evaluated from the

Table 3.1 Contributions of GVA accounts to T and C

Account code	Contribution to L	Contribution to T	Contribution to C
6	*6*	–	–
7a	–	$(k_a + (1 - k_a).TI).7a$	$(1 - k_a).(1 - TI).7a$
7c	–	$TI.k_p.7c$	$(1.TI.k_p).7c$
Y	–	$TI.k_p.Y$	$(1 - TI.k_p).Y$
POA	–	$TI.k_p.POA$	$(1 - TI.k_p).POA$

yearly depreciation map. A higher k_a means that the production process uses a higher value from TA relative to the use of total assets.

The GVA coefficient k_p indicates how much GVA weights relative to ordinary activity total costs, $k_p = GVA/(GVA + IPV)$. A higher k_p indicates lower costs with intermediate consumptions, such that a higher value of current assets is used for the formation of the GVA.

Using appropriately the technological index TI, the technology coefficient k_a and the GVA coefficient k_p, it is now possible to identify the parts of each of the accounts that builds the GVA identity (3.3) as contributing to the terms T and $\underline{C}$. This is shown in Table 3.1.

Finally, the equations for evaluating L, T, and $\underline{C}$ are shown below where the account code numbers referenced above are written in italics. The sum of $L + T + \underline{C}$ is the KTC identity (3.1) and (3.7).

$$L = 6 \tag{3.4}$$

$$T = GVA.\,k_a.\,7a/\left(GVA - (1 - k_a).\,7a - k_p.\,(17 + 7c + 13 + Y + 21 - 16 - 9 \text{ to } 11)\right) \tag{3.5}$$

$$C = (1 - k_a).\,7a.\,(1 - TI) + \left(1 - k_p.\,TI\right).\,(17 + 7c + 13 + Y + 21 - 16 - 9 \text{ to } 11) \tag{3.6}$$

$$L + T + \underline{C} = GVA \tag{3.7}$$

To compute the technological coefficient k_a, as described above, it is necessary to know the depreciation value for each type of asset, which is the case for the gathered Portuguese data to be presented in Chap. 4. However, the European database BACH does not have that data. An alternative algorithm for computing the technological coefficient k_a was then developed considering the assets' values instead of the assets' depreciation values, and then correcting according to an empirically calculated factor fe. This is done as follows: The technological coefficient k_a is computed by the ratio $A/(A + B)$, where A is the depreciation value of equipment, tools, industrial property plus R&D costs and B is the depreciation values of all other fixed assets. The TA, using BACH's Balance Sheet codes, are named (*C.1.5* + *C.2.2* + *C.2.3*), and all the other fixed assets $\underline{CA}$ are named (*C.1.1* + *C.2.1* + *C.2.4*). Considering the ratio A/TA as the average depreciation coefficient of technological forms δ_a, and $B/\underline{CA}$ as the average depreciation coefficient of all the other fixed assets δ_b, the ratio $k_a = A/(A + B)$ may be written as fe.TA/(fe.TA + CA), where fe $= \delta_a/\delta_b$. The alternative algorithm to compute k_a is written as in the following identity.

$$k_a = \text{fe.}(C.1.5 + C.2.2 + C.2.3) / [\text{fe.}(C.1.5 + C.2.2 + C.2.3) + (C.1.1 + C.2.1 + C.2.4)] \tag{3.8}$$

The empirical factor fe corrects the different depreciation coefficients that are used for different asset types. For example, a typical depreciation coefficient for buildings is 0.02–0.04, but for computer equipment is 0.33. These are extreme examples, but, on average, TA have depreciation coefficients two to three times higher than capital assets'. The empirical factor fe depends on the balance between technology and capital assets, and thus it changes with the sector under analysis. As an example, comparing the Portuguese data from INE[1] with the corresponding data from BACH, it was empirically calculated for the manufacturing sector, giving fe = 2.6.

3.4.1 How the Algorithm Works

To better understand the results of this algorithm and how profit and production costs influence the three GVA contributions L, T, and C, I take a fictional firm or sector described here by its Profit and Loss Account, shown in Table 3.2.

The calculation of the technological coefficient k_a also requires knowing how different types of assets depreciate and the respective depreciation values during the time period under analysis. We shall posit the values shown on Table 3.3. Assuming that "Other intangible assets" refer to Industrial property and that "Other fixtures" refer to R&D expenses, k_a value equals (12 + 5 + 5)/30 = 0.73.

The GVA coefficient is computed as explained before, k_p = GVA/(GVA + IPV), where the IPV is the difference between the ordinary activity total costs and GVA. According to Table 3.2, the ordinary activity total costs are the sum of the accounts ($5 + 8 + 6 + 7 + 13 + Y$). As such, k_p = 100/170 = 0.59.

According to expressions (3.4–3.6), the value contributions from the uses of labor L, of technology T, and of capital C are shown in Table 3.4. We may conclude that, from a GVA of 1,000 units of value, 550 originate from the use of knowledge, 264 from the use of technology and 186 from the use of capital. The respective indexes are approximately 55, 26, and 19 %.

To better understand how the value contribution and their indexes change with traditional production parameters, the three following exercises will be described next, where the changes refer to the initial reference shown in Tables 3.2, 3.3, and 3.4.

[1] INE—Instituto Nacional de Estatística (National Statistics Institute).

Table 3.2 Profit and loss account of a fictional firm

Code	Profit and loss account	u.v.
1	Net turnover	1,450
2	Change in stock...	100
3	Capitalised production	150
4	Other operating income	50
S	*Total operating income (S = 1 + 2 + 3 + 4)*	1,750
5	Costs of materials and consumables	650
5a	Raw materials and consumables	650
5b	Other external charges	0
8	Other operating charges and taxes	100
T	*Added Value BACH (S − 5 − 8)*	1,000
6	Staff costs	550
6a	Wages and salaries	450
6b	Social security costs	100
U	*Gross operating profit (T − 6)*	450
7	Value adjustments on non-financial assets	300
7a	Depreciation on intangible and tangible assets	300
7c	Other value adjustments and provisions	0
V	*Net operating profit (U − 7)*	150
9/11	Financial income	150
12	Value adjustments on financial assets	0
13	Interests and similar charges	50
13a	Interest paid on financial debts	50
13b	Other financial charges	0
W	*Financial income net of charges (9/11 − 12 − 13)*	100
X	*Profit on ordinary activity before taxes (V + W)*	250
16	Extraordinary income	100
17	Extraordinary charges	50
Y	Taxes on profit	50
21	*Profit or loss for the financial year (X + 16 − 17 − Y)*	250

Table 3.3 Depreciation values according to different types of fixed assets (fictional firm)

		Assets value	Annual depreciation value
C	*Fixed assets*	2,300	300
C.1	*Intangible fixed assets*	350	60
C.1.1	Formation (preliminary) expenses	100	10
C.1.5	Other intangible fixed assets	250	50
C.2	*Tangible fixed assets*	1,950	240
C.2.1	Land and buildings	1,000	50
C.2.2	Plant and machinery	600	120
C.2.3	Other fixtures	250	50
C.2.4	Payments on account and assets in construction	100	20

Table 3.4 Use values of knowledge *L*, of technology *T* and of capital C

Value contributions	u.v.	Indexes	Percentage (%)
L	550	KI	55
T	264	TI	26
C	186	CI	19
GVA	1,000		100

a. To increase Net turnover by 20 %, from 1,450 to 1,740, by only increasing prices, *ceteris paribus*. It results a 116 % increase in profit on ordinary activity before taxes and a 29 % increase in the GVA. C shows a 124 % increase, T increases by 23 % and L remains unchanged. The indexes are shown in Table 3.5.
b. Increasing the relative weight of technology assets, keeping constant total depreciation: Plant and machinery depreciation up from 120 to 150, and buildings depreciation down from 50 to 20, *ceteris paribus*. The results do not change in both GVA and profit of ordinary activity before taxes. C decrease to 86 % of the reference value, T increases by 10 % and L remains unchanged. The indexes are shown in Table 3.6.
c. Decreasing Interests and similar charges by 100 %, from 50 to 0, *ceteris paribus*. There results no change in GVA and the profit of ordinary activity before taxes increases the same amount. As such, C slightly decreases, T slightly increases, and L remains unchanged. The indexes are shown in Table 3.7.

Table 3.5 Use values of knowledge L, of technology T, and of capital C

Value contributions	u.v.	Indexes	Percentage (%)
L	550	KI	43
T	324	TI	25
C	416	CI	32
GVA	1,290		100

Table 3.6 Use values of knowledge L, of technology T, and of capital C

Value contributions	u.v.	Indexes	Percentage (%)
L	550	KI	55
T	290	TI	29
C	160	CI	16
GVA	1,000		100

Table 3.7 Use values of knowledge L, of technology T, and of capital C

Value contributions	u.v.	Indexes	Percentage (%)
L	550	KI	55
T	265	TI	27
C	185	CI	18
GVA	1,000		100

3.5 KTC Dynamic Model and Growth Conditions

The KTC model's fundamental identity is the expression (3.1), rewritten below substituting GVA by *Y* in (3.9). This is a static identity. In order to reveal how the parameters may change with time, it will be written the respective dynamic equation. Considering partial derivatives to time t and writing them with a dot on top of the respective symbol, we may write (3.10). Dividing it by *Y*, it is possible to describe the model's dynamics in terms of growth rates (3.11). For example, the GVA growth rate is $\widehat{y} = \dot{Y}/Y$. This shows that the GVA growth rate $\widehat{y}$ equals the sum of growth rates of the three parameters L, T, and $\underline{\mathrm{C}}$, each one multiplied by the respective index KI, TI, and $\underline{\mathrm{CI}}$.

$$Y = \mathrm{L} + \mathrm{T} + \underline{\mathrm{C}} \tag{3.9}$$

$$\dot{Y} = \dot{\mathrm{L}} + \dot{\mathrm{T}} + \dot{\underline{\mathrm{C}}} \tag{3.10}$$

$$\widehat{y} = \widehat{\ell}.\,\mathrm{KI} + \widehat{t}.\,\mathrm{TI} + \widehat{\underline{c}}.\,\mathrm{CI} \tag{3.11}$$

One other way of writing the dynamic growth conditions uses the ratio *Y* over labor costs L, which I name knowledge productivity KP, and measures how human knowledge (work value) may directly produce added value. Knowledge productivity describes the value added per unit of labor cost, and is the inverse of what traditionally is referred to as unit labor cost, which is the cost of labor per unit of GVA. An increase of KP means increases of T and/or $\underline{\mathrm{C}}$ in face of L, in other words, the knowledge productivity increase means additional technology and capital use value with a constant labor value. In terms of growth rates, (3.11) can be written as (3.13).

$$\mathrm{KP} = Y/\mathrm{L} = 1 + \mathrm{T}/\mathrm{L} + \underline{\mathrm{C}}/\mathrm{L} = 1/\mathrm{KI} \tag{3.12}$$

$$\widehat{y} = \widehat{\ell} + \widehat{\mathrm{k}}_{\mathrm{p}} \tag{3.13}$$

This equation allows a very straightforward and clear conclusion on economies growth conditions. As L and KP are not independent, for *Y* to increase at a growth rate $\widehat{y}$, both L and KP must increase, or at least one should increase and the other remaining constant. In other words, it is not enough to increase L because that might decrease KP, and it is not enough to increase KP, because that may originate in a decrease of L. To better understand this growth law, we should remember that the value added also equals total expense, and expense equals consumption plus investment (saving). In a production environment, increasing L implies less value available for investment and then less likely to increase KP, the inverse being equally true.

A simple example showing three situations illustrates this effect. Taking again the values shown in Tables 3.2 and 3.3 as a reference, we shall see what happens to KP and Y when changing L, T, and $\underline{C}$.

1. Reducing L by 20 %, from 550 to 440. The immediate result is an increase in net operating profits from 150 to 260, with the GVA constant. Subsequently, T and $\underline{C}$ increase, what makes KP to increase from 1.82 to 2.27. These numbers are shown in Table 3.8. Checking expression (3.13), $\widehat{\ell}$ in negative, $\widehat{k_p}$ is positive and the growth rate $\widehat{y}$ is zero.
2. Reducing depreciation 20 % (from 300 to 240). The immediate result is an increase in net operating profits from 150 to 210, keeping GVA constant. Subsequently, T + $\underline{C}$ remain equal, what makes KP to keep the same value 1.82. These numbers are shown in Table 3.9. Checking expression (3.13), $\widehat{\ell}$ and $\widehat{k_p}$ are zero and so is $\widehat{y}$.
3. Increasing net turnover by 20 %, from 1,450 to 1,740, *ceteris paribus*, only by increasing prices. The immediate result is an increase in profit on ordinary activity before taxes from 250 to 540, and an increase in the GVA from 1,000 to 1,290. Subsequently, T + $\underline{C}$ increase from 450 to 740 and KP increases from 1.82 to 2.35. These numbers are shown in Table 3.10. Checking expression (3.13), $\widehat{\ell}$ is zero, $\widehat{k_p} = 0.29$ and so is $\widehat{y} = 0.29$.

Table 3.8 Use values of L, T, and $\underline{C}$, and the value of knowledge productivity KP

Value contributions	u.v.	Indexes	Percentage (%)
L	440	KI	44
T	291	TI	29
$\underline{C}$	269	$\underline{CI}$	27
GVA	1,000		100
KP	2.27		

Table 3.9 Use values of L, T, and $\underline{C}$, and the value of knowledge productivity KP

Value contributions	u.v.	Indexes	Percentage (%)
L	550	KI	55
T	212	TI	21
$\underline{C}$	238	$\underline{CI}$	24
GVA	1,000		100
KP	1.82		

Table 3.10 Use values of L, T, and $\underline{C}$, and the value of knowledge productivity KP

Value contributions	u.v.	Indexes	Percentage (%)
L	550	KI	55
T	324	TI	25
$\underline{C}$	416	$\underline{CI}$	32
GVA	1,290		100
KP	2.35		

Concluding, this growth law tells that economies grow when both L and KP increase or, at least, one increase and the other remains constant. The relative increases of L and KP, although resulting in immediate growth, will depend on particular situations and have different implications: (1) When L increases and KP remains constant, profit decreases, and thus there will be less cash available for investment and more cash available for consumption, what may result in higher prices; (2) When KP increases by the increase of profits and L remains constant, there will be more cash available for investment and the same cash for consumption, what may result in lower prices. The optimum seems to correspond to increases in both, sometimes higher for KP, sometimes higher for L. Increases of both L and KP imply, in conclusion, an increase in investment. Following this line of thought, a model can be built to find the optimum growth path during a certain period of time [4].

References

1. Aken JE (2004) Management research based on the paradigm of the design sciences: the quest for field-tested and grounded technological rules. J Manag Stud 41(2):219–246
2. BACH Database, EUROSTAT, Bank for the Accounts of Companies Harmonized. Guide for Database Users—4.6—Profit and Loss Account. http://ec.europa.eu/economy_finance/indicators/bachdatabase/bachdatabase_whatisbach_en.htm
3. Chu D (2008) Criteria for conceptual and operational notions of complexity. Artif Life 14(3):313–323
4. Cruz DF (2012) A growth strategy model—growth analysis for firms and sectors. Master Thesis in Engineering and Industrial Management, Departament of Engineering and Management, Instituto Superior Técnico—UTL, Lisbon, Portugal
5. Edward MR (2005) Organizational identification: a conceptual and operational review. Int J Manag Rev 7(4):207–230
6. ESA 1995 (1996) European system of accounts. Office for Official Publications of the European Communities, Luxembourg
7. Fernandes ASC (2012) Assessing the technology contribution to value added. Technol Forecast Soc Change 79:281–297
8. Morrow P (1983) Concept redundancy in organizational research: the case of work commitment. Acad Manag Rev 8(3):486–500
9. Venkatraman N, Ramanujam V (1987) Planning system success: a conceptualization and an operational model. Manag Sci 33(6):687–705

Chapter 4
The Value Added by Technology

4.1 Comparing Economic Activity Sectors

Following Fernandes [1], the proposed KTC model will now be applied, computing the knowledge, technology, and capital indexes of different economic sectors and divisions within one economy. The Portuguese economy was chosen because all the necessary data were readily available from INE, including depreciation values per different types of assets. This economy will be characterized in what concerns its sectors' dependence on knowledge, technology, and capital by computing the values added by the uses of knowledge, technology, and capital. For comparing economic activity sectors, the algorithm proposed in Chap. 3 was applied using data from years 1996 to 2003. This analysis considered the universe of firms with more than 20 employees for 49 divisions of CAE[1] rev. 2 (NACE 1.1 or ISIC rev. 3). The divisions that are not covered in this study, because reliable data were not available, are the following: In the secondary sector, divisions 10 Mining of Coal and 11 Extraction of gas and petroleum; in the Tertiary, are not considered one large division (75 public administration), four very small divisions (73 Research & development, 91 Activities and membership organizations NEC; 95 private households; and 99 extra-territorial organizations), and the financial activity (65 financial intermediation; 66 Insurance; 67 activities auxiliary to financial). The data and financial maps needed for this study are the following: The profit and loss account, from where the standard identity for the GVA can be computed; the balance sheet, with both gross and net values of assets and accumulated depreciations; the depreciation map, with detailed depreciation values per each type of assets.

For the year 2000, the results for indexes KI, TI, and CI are shown in Fig. 4.1 for the primary (divisions 1–5), secondary (10–45, except 10 and 11), the tertiary (50–99, except 65–67, 73, 75, 91, 95 and 99) and the sum of the three, named Total.

---

[1] Classificação de Actividade Económica (Classification of Economic Activity).

A. S. C. Fernandes, *The Contribution of Technology to Added Value*,
DOI: 10.1007/978-1-4471-5001-5_4, © Springer-Verlag London 2013

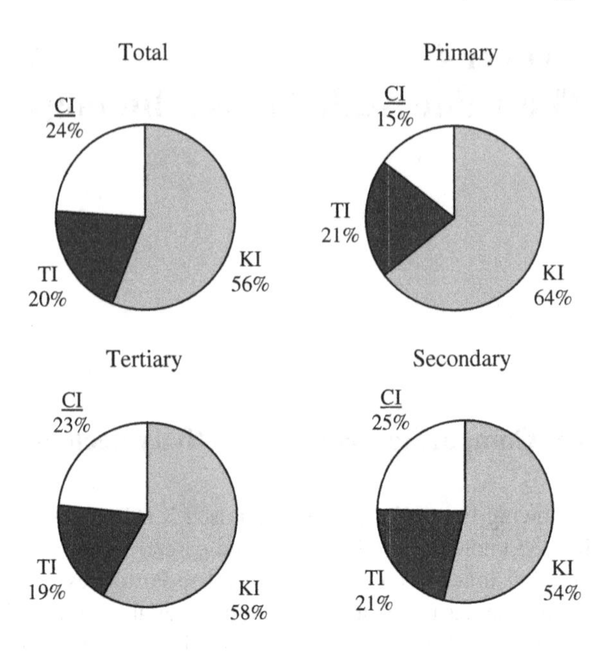

Fig. 4.1 Knowledge, technology, and capital indexes. Primary (NACE 1.1 divisions 1–5), secondary (10–45, except 10), tertiary (50–99, except 65–67, 75, 91, 95, and 99) and total (year 2000, universe of Portuguese firms with more than 20 employees) (Reprinted from Fernandes [1], Copyright (2012) reproduced with permission from Elsevier)

The results show that, for the total activity, the knowledge index is 56 %, whereas the technology index is 20 % and the capital index is 24 %. The direct contribution of knowledge accounted for more than half the total GVA and almost two-third in the primary sector, what makes this sector highly dependent on labor and thus on the direct use of knowledge. The TI is not very different in the three sectors. Still, the primary and the secondary use relatively more technology that the tertiary. As for CI, the highest value is found in the secondary, where the capital borrowed, the buildings' depreciation values and profits seem to be higher. The indexes evolved within this period as shown in Figs. 4.2, 4.3, and 4.4, a period that was characterized as an expansion from 1996 to 2000 and a recession up to 2003.

Next, results for a few sectors are described in Fig. 4.5, indexes are shown for section D (Manufacturing–divisions 15–37), division 32 (Manufacturing radio, etc.), division 40 (Electricity), and division 45 (Construction). It is interesting to compare the high TI and CI of electricity, which are, respectively, 34 and 42 %, with the much lower ones of construction, which are, respectively, 15 and 19 %. Electricity is commonly known as a capital-intensive sector, which is corroborated by these results, and construction, as expected, shows a high dependence on labor.

In Fig. 4.6, a direct comparison is shown among four very different divisions: Retail trade (52), telecommunications (642), computer and related activities (72), and education (80). The KI is very high for education and high for computers, showing their high dependence on knowledge and labor. In retail, KI is close to the country's average and in telecommunications is very low. On the other hand, the TI is high in telecommunications and very low in education and in computers. It is surprising to find out that the division computer and… has a very low

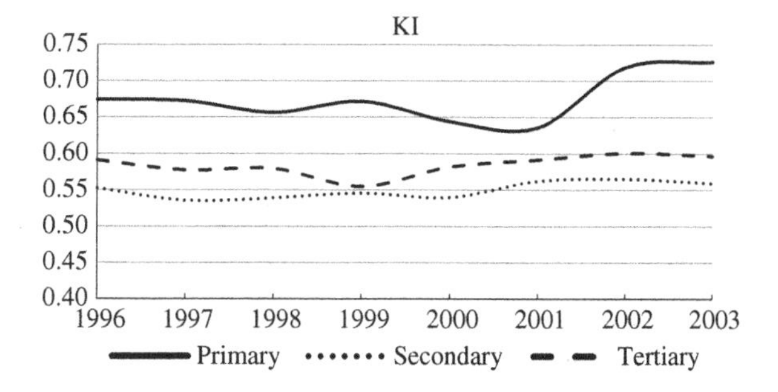

Fig. 4.2 Portugal knowledge index

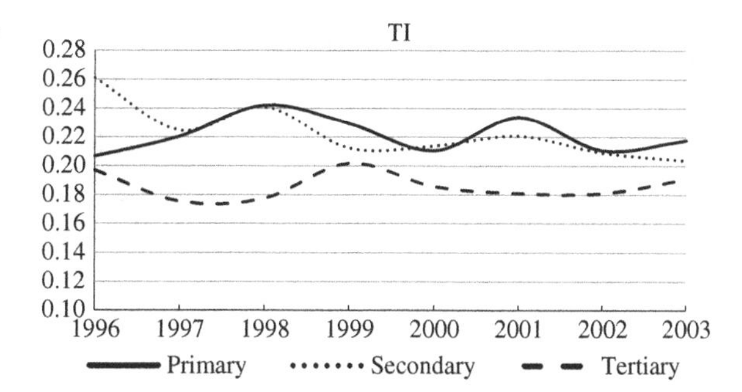

Fig. 4.3 Portugal technology index

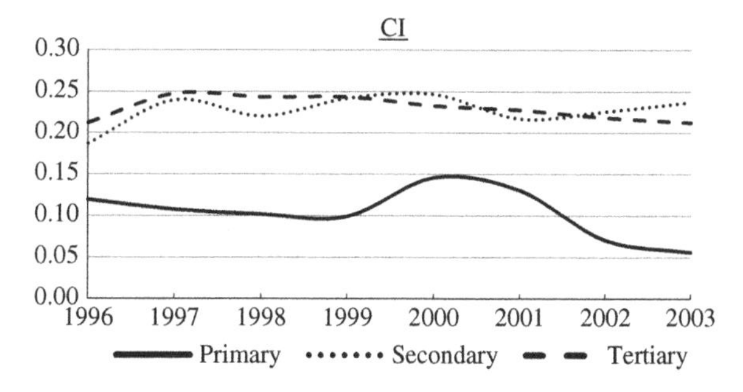

Fig. 4.4 Portugal capital index

technology index, 8 %, which is almost as low as education's 6 %. They are both knowledge-intensive divisions, very far from being high-technology divisions. When comparing divisions like computer and related activities with construction, where KI shows that labor costs are in both the major contribution to GVA, one may be lead to the idea that these divisions have similar characteristics. However, that is not the case, because the former has a number of workers highly paid and the latter has a much larger number of workers but at a much lower average wage. This effect calls for attention to complement the information given by these indexes with other indicators, such as labor productivity.

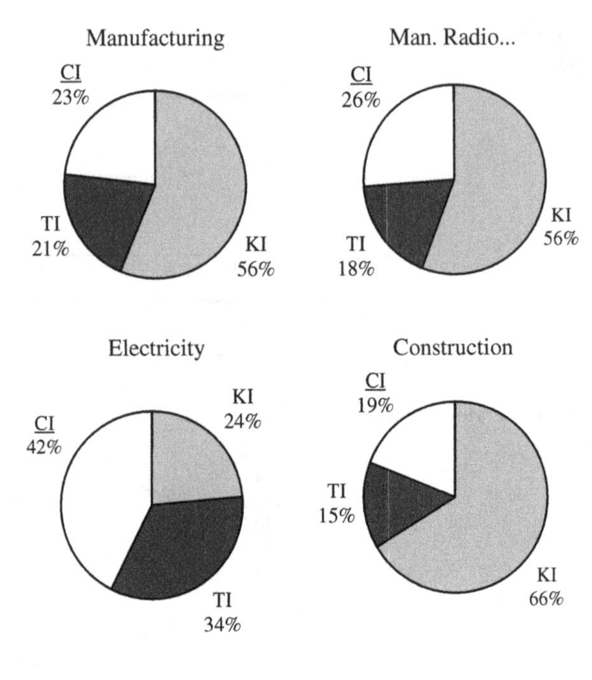

Fig. 4.5 Knowledge, technology, and capital indexes: whole of manufacturing industry (divisions 15–37); to manufacturing of radio, television, and communication equipment (32); electricity… (40); and construction (45) (year 2000, Portuguese firms with more than 20 employees) (Reprinted from Fernandes [1], Copyright (2012) reproduced with permission from Elsevier)

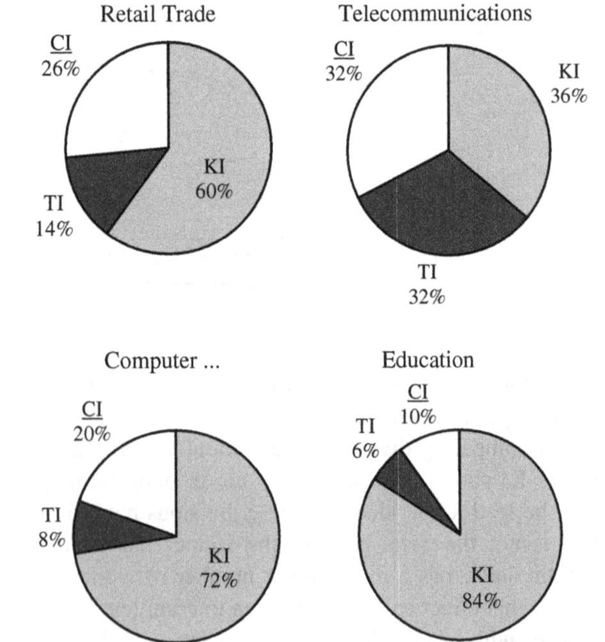

Fig. 4.6 Knowledge, technology, and capital indexes of divisions retail trade (52), telecommunications (642), computer and related activities (72), and education (80) (year 2000, universe of Portuguese firms with more than 20 employees) (Reprinted from Fernandes [1], Copyright (2012) reproduced with permission from Elsevier)

4.2 Comparing Economies

The values L, T, and $\underline{C}$, as well as their respective indexes, KI, TI, and $\underline{CI}$, were also computed for the manufacturing sector (NACE Rev 1.1) of all the European countries that have the relevant data available in the European communities database BACH: Portugal, Belgium, Germany, Spain, France, Finland, and Poland. All sizes of firms were considered, with the variable sample choice and data spanning the years 1995–2008. The algorithm used in this analysis for computing L, T, and $\underline{C}$ had a small change compared with the one used for Portuguese data. The reason is that there are no depreciation map accounts in this database. This slightly different and less accurate algorithm builds on what was learned from the algorithm's implementation for the Portuguese data. This was explained in the last part of Chap. 3, Sect. 3.4.

For the manufacturing sector, the results for the indexes KI, TI, and $\underline{CI}$ are shown in Table 4.1, for the three periods 1995–2000, 2001–2006, and 2007–2008. The values for each period are averages of 6 years' annual index percentage values, for the two initial periods, and an average of 2 years' annual index percentage values, for the last period. Results show different indexes for different countries and important changes along the three periods. The most striking evidence is Germany's very high manufacturing industry dependence on labor contribution to value added (KI). Germany and France are the countries where TI is lower and KI is higher. Economies with lower relative dependence on knowledge have, consequently, higher relative dependences on technology and $\underline{\text{capital}}$. $\underline{CI}$ is always higher than TI, especially in less developed economies where investment is typically lower.

Along the period 1995–2008, which covers approximately one business cycle, we may note the following:

- KI: Belgium, Germany, and France decrease their Knowledge Index (the same as unit labour cost), while Portugal, Spain, Finland, and Poland increase it. This,

Table 4.1 Values of indexes KI, TI and $\underline{CI}$ for different European countries' manufacturing sectors (1995–2008)

	Portugal	Belgium	Germany	Spain	France	Finland	Poland
Average 1995–2000							
KI	0.543	0.628	0.747	0.599	0.660	0.495	0.543
TI	0.209	0.184	0.112	0.138	0.107	0.120	0.209
$\underline{CI}$	0.243	0.187	0.127	0.252	0.215	0.378	0.243
Average 2001–2006							
KI	0.565	0.615	0.742	0.620	0.653	0.537	0.496
TI	0.182	0.169	0.118	0.139	0.102	0.129	0.130
$\underline{CI}$	0.246	0.216	0.121	0.216	0.224	0.313	0.368
Average 2007–2008							
KI	0.598	0.617	0.708	0.637	0.657		0.493
TI	0.147	0.158	0.122	0.133	0.087		0.118
$\underline{CI}$	0.245	0.223	0.144	0.214	0.236		0.372

for the first group of countries, indicates policies giving priority to investment rather than distribution of value.

- TI: Germany and Finland increase their technology index, while the other countries decrease it. This shows that in the two countries it was given priority to investment in technology.
- CI: Portugal, Belgium, Germany, France, and Poland increase their capital index, while the other countries decrease it.

There is a good evidence of the Germany economy's robustness in 2008, as it succeeded, along the previous business cycle, to decrease its unit labor cost (ULC equals KI) and increase both the technology index and the capital index, and thus becoming better prepared for the incoming financial crisis of 2008–2009. From this group, it was the only country where this happened.

At the beginning of this crisis the indexes changed as shown in Table 4.2, for the years 2009 and 2010. All countries show a decrease in the capital index (CI), although Germany shows a very small decrease. The decrease is due to mainly the fall in profits. This index decrease implies an increase on the other two indexes. The knowledge index KI increases in all countries, except for Poland where it remained almost constant. The technology index also increased in all countries, except for Germany where it remained almost constant. The indexes showed in Tables 4.1 and 4.2 are described in Figs. 4.7, 4.8, and 4.9.

Table 4.2 Values of indexes KI, TI, and CI for different European countries' manufacturing sectors (2009–2010)

	Portugal	Belgium	Germany	Spain	France	Poland
Average 2009–2010						
KI	0.625	0.627	0.720	0.697	0.688	0.497
TI	0.153	0.171	0.121	0.156	0.095	0.157
CI	0.217	0.203	0.140	0.143	0.192	0.338

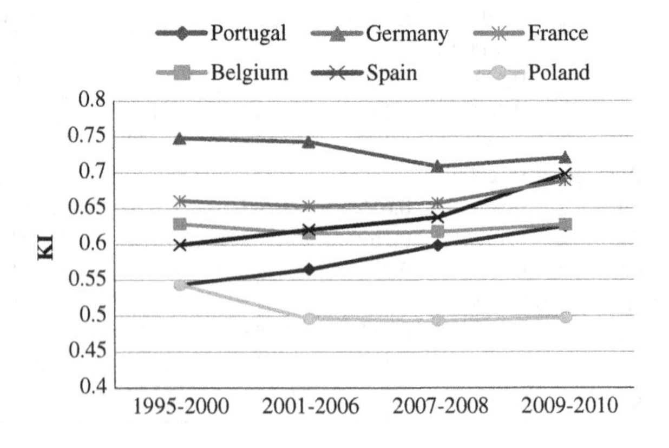

Fig. 4.7 Knowledge index KI for different European countries' manufacturing sectors

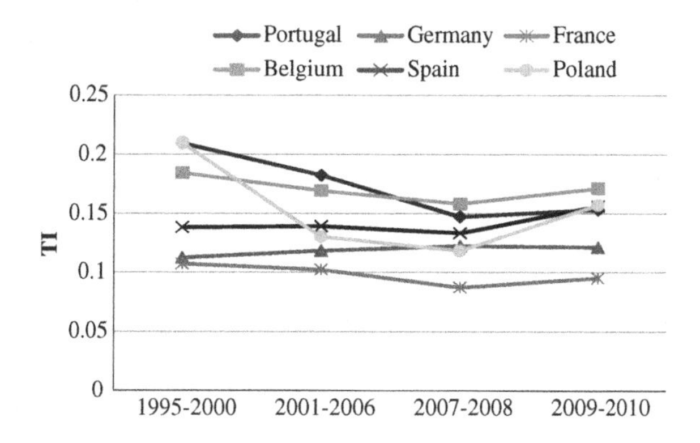

Fig. 4.8 Technology index TI for different European countries' manufacturing sectors

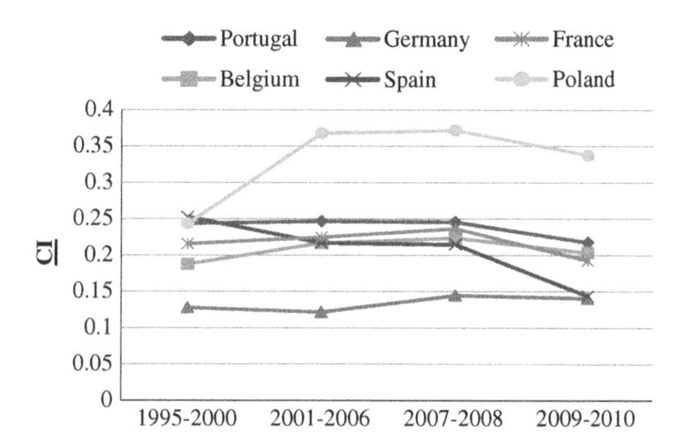

Fig. 4.9 Capital index CI for different European countries' manufacturing sectors

4.3 Conclusions

This chapter shows results of applying the KTC model algorithms to find out how much technology, knowledge, and capital contribute to GVA. These results help to understand the nature of the indexes TI, KI, and CI and how they can help to characterize economy changes and growth. As for any other indexes, they add a new angle through which understanding the whole picture becomes more complete. As they are objectively linked to international standard accounting, it is easier to, through them, describe objectively the technology dependence of a sector, a firm or a whole economy and compare them independently of the year or of the country. There is complementary information that would be important to compute, like the values of used technology per hour worked or per employee, and equally for capital.

Some results are especially important because they challenge some current thoughts and beliefs. A first example is the similar technology index (TI) of sectors like computer and… (72) and education (80). They are both knowledge intensive sectors where labor represents the majority of value contribution to GVA.

Even if R&D and innovation is far greater in the former, the real activity and value added show clearly that the highest dependence is from knowledge, expressed as work, and valued by the corresponding labor. A second unexpected result is finding about the same knowledge dependence on sectors like computer and... (72) and construction (45). The former has fewer employees but each with a larger salary, such that the final picture looks about the same. Finally, a third striking conclusion is that countries like Germany and France, when compared with Portugal, Belgium, Finland, and Poland, show a much lower technology index in their manufacturing sector. At the same time, those two countries show a larger knowledge index. This result shows how these indexes depend on the value distribution policies of the countries. In fact, German and French firms pay their labor higher wages, guaranteeing their competiveness through higher productivities.

Reference

1. Fernandes ASC (2012) Assessing the technology contribution to value added. Technol Forecast Soc Chang 79:281–297

Chapter 5
Technology Dependence Taxonomy

The technology index (TI) relates the value added by the use of technology T to total GVA. I advocate that it represents the firm's or sector's dependence on technology better than the parameter. TI that currently is used to classify economic activity sectors as of high or low technology dependence. TI is a factor used by OECD to designate how much an industrial sector depends on technology; though it exclusively specifies its dependence on R&D, because it is computed as the ratio between R&D costs and GVA. In this way, the TI parameter shows the technological innovation effort but it does not show the whole dependence on technology use. Also, it does not hold its significance when comparing the same sectors in different economies because they may have very different R&D efforts. Taking as an example the division 32—manufacture of radio, television, and communication equipment: To this division it was attributed a TI of 18.65 % in the USA [3] whereas in Portugal, the average of the TI from 1996 to 2003 was almost zero [2]. Accordingly, one would conclude that this division is a high technology division in the USA and a low-technology one in Portugal. To further illustrate this problematic approach, Table 5.1 shows the OECD's technological intensity compared with the TI from the KTC model, for various industrial sectors, where the difference of the two criteria is pointed out.

Divisions 32, 31, 26, and 15 + 16 in the USA are classified on their technology dependence, according to OECD criterion, as of high technology, medium–high technology, medium–low technology, and low technology, respectively. In Portugal, the same divisions show zero technology dependence, while, if classified according to the TI, they show an effective technology dependence of 18, 16, 29, and 23 %, respectively.

Such a different classification of technology dependence of the same sectors in two countries indicates that the metric is not adjusted to the goal.

A. S. C. Fernandes, *The Contribution of Technology to Added Value*,
DOI: 10.1007/978-1-4471-5001-5_5, © Springer-Verlag London 2013

Table 5.1 **a** Technology intensities and OECD's classification of several industrial sectors in the USA. **b** For the same industrial sectors, in Portugal, the second column indicates what would be the OECD's classification and the TI computed by the KTC model

a

Industrial sectors, USA–ISIC rev.3	OECD's technology intensity (%)	OECD's technology dependence classification (1990)
Manufacture of radio, TV and communication … (division 32)	18.65	High technology
Manufacture of electric machinery … (division 31)	7.63	Medium–high technology
Manufacture of other non-metallic minerals … (division 26)	2.2	Medium–low technology
Manufacture of food products and beverages and tobacco products (division 15 + 16)	1.14	Low technology

b

Industrial sectors, Portugal—CAE rev.2	OECD's technology intensity (%)	Technology index (TI) (1996 a 2000) (%)
Manufacture of radio, TV and communication … (division 32)	0 (Low technology)	18
Manufacture of electric machinery … (division 31)	0 (Low technology)	16
Manufacture of other non-metallic minerals (division 26)	0 (Low technology)	29
Manufacture of food products and beverages and tobacco products (division 15 + 16)	0 (Low technology)	23

5.1 A New Metric for a New Taxonomy

Establishing a metric for sectors' classification concerning technology dependence becomes easier to propose if considering the TI statistical distribution in an economy. We shall start by looking at how the three indexes relate to each other in a real economy in order to better understand their distributions.

The three indexes KI, TI, and $\underline{\text{CI}}$ must add to unity and so are not independent. The values of KI, TI, and $\underline{\text{CI}}$ were computed for 49 Portuguese divisions along 8 years, 1996–2003. In a three-dimensional space $\{\text{KI}, \text{TI}, \underline{\text{CI}}\}$ the points are over a plane surface, as it may be observed in Fig. 5.1 where the results for the year 2000 are plotted. In this space, one point representing division i has the coordinates $\{\text{KI}_i, \text{TI}_i, \underline{\text{CI}}_i\}$. The plane surface equation is $\text{KI} = 1 - \text{TI} - \underline{\text{CI}}$.

To examine how the indexes depend on each other, the same group of points may be seen in a two-dimensional space, projecting them in one face of the above 3D space, first on the $\{\text{KI}, \underline{\text{CI}}\}$ face, second on the $\{\text{KI}, \text{TI}\}$ face and third on the

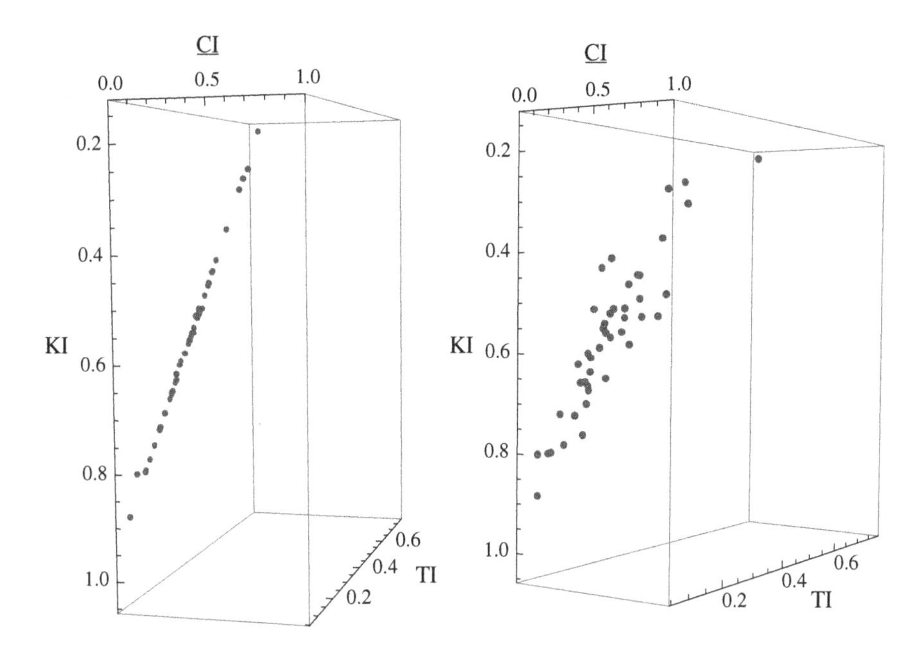

Fig. 5.1 Two visualizations of the 3D space {KI, TI, CI}, where each point represents one division (49 divisions of the Portuguese economy, year 2000). On the *left*, the visualization angle was chosen such that all points can be seen on a unique plane surface

{CI, TI} face. This is depicted in Fig. 5.2,[1] where correlations were analyzed between the pairs of indexes {KI, CI}, {KI, TI}, and {CI, TI}. The point distributions on the three planes show weak correlations as seen by the low correlation factors R^2. These correlations, even if weak, are explained as follows: The parameters *L*, T, and C were assumed to be independent, such that one of them may be changed without affecting the others. For example, a firm may decide to raise its labor costs (higher *L*) or build new headquarters (higher C) without affecting, respectively, T and C in the first case and *L* and T, in the second case. However, the respective indexes are not independent because they must sum to the unity. As such, when one index changes the others change in opposite direction; as for example, for a higher *L*, *ceteris paribus*, KI increases and, necessarily, TI + CI decreases. But does each index decreases at the same rate? The plane {KI, CI} shows values for KI with typically twice the values of CI, and a fair negative correlation represented by the factor $R^2 = 0.55$. On the plane {KI, TI}, the same effect can be observed but the correlation is weaker, $R^2 = 0.33$. On the plane {CI, TI}, both indexes have values under 0, 5 and the correlation is absent. The overall conclusion is that a higher KI implies necessarily a smaller TI + CI and it

[1] Reprinted from Publication Technological Forecasting and Social Change 79, António S C Fernandes, Assessing the Technology Contribution to Value Added, Copyright (2012), 281-297, with permission from Elsevier.

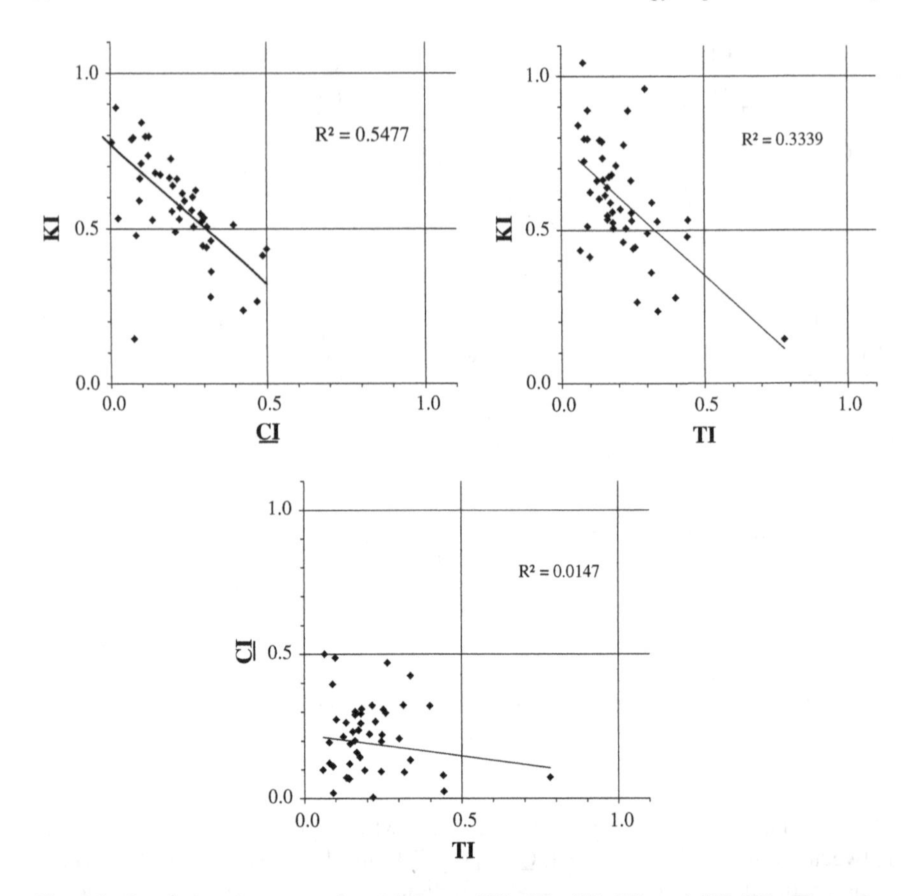

Fig. 5.2 Correlations between pairs of indexes, {KI, CI}, {KI, TI}, and {CI, TI}. (Fernandes [1], reproduced with permission from Elsevier

correlates better with a smaller C than with a smaller T, what is easy to understand as higher labor costs imply, in most cases and in the short term, lower profits, sinking the value of C and the value of CI.

I propose a metric for technology dependence of firms, sectors, and economies according to a criterion and a reference. The criterion is the TI, at four levels: Low, medium–low, medium–high, and high technology dependence; and the reference values are the four quartiles of the TI distribution in a reference economy to be chosen. In the case of the above data, distributions of the three indexes are shown in Table 5.2.

Accordingly, the sectors classification in what relates to technology dependence would be as shown in Table 5.3. It is important to note that the TI distribution is somehow dependent on the economy under study. The main reason is that every economy distributes its value gains by salaries or in savings (investment) in different ways, depending on the respective economic development situation. Still, those differences are relatively small as it was verified in Table 4.2.

Table 5.2 Mean, median, standard deviation, and percentiles for indexes KI, TI, and CI (Portugal 1996–2003 years)

Index	Mean	Median	Standard deviation	Percentile				
				0 %	25 %	50 %	75 %	100 %
KI	0.601	0.603	0.180	0.125	0.503	0.603	0.710	1.180
TI	0.218	0.182	0.140	0.016	0.129	0.182	0.271	0.886
CI	0.185	0.194	0.153	−0.369	0.103	0.194	0.277	0.586

Table 5.3 Proposed criteria for sector's technology dependence classification

TI distribution quartile	TI	Classification
First quartile	<0.13	Low technology
Second quartile	0.13–0.19	Medium–low technology
Third quartile	0.20–0.27	Medium–high technology
Fourth quartile	>0.27	High technology

Table 5.4 Examples of sector's technology dependence classification, according to proposed criteria

Division	TI (%)	Technology dependence
Manufacturing of radio … (32)	15	Medium–low
Electricity (40)	34	High
Construction (45)	15	Medium–low
Retail trade (52)	13	Medium–low
Telecommunications (642)	32	High
Computer and related … (72)	8	Low
Education (80)	6	Low

According to this criterion and taking the values shown on Figs. 4.5 and 4.6 as examples, the sectors would be classified as shown in Table 5.4.

5.2 Technological Content of a Product

The last section was devoted to analyze the technology contribution to the GVA of a firm or a sector. The same method may be used to evaluate the technological content of a product. Let us consider a hypothetical firm, which output comprises one only product, as depicted in Fig. 3.2. Its final product value (FPV) equals the sum of the GVA with the intermediate product value (IPV). The value chain of this product could be represented as shown in Fig. 5.3. The IPV of that firm IPV_1 is the sum of the FPVs of a number of firms belonging to a second level in the value chain, which, in turn, are the sum of the respective GVA and IPV. This repeats along the levels of the chain up to infinity, both in what concerns levels' firms and time.

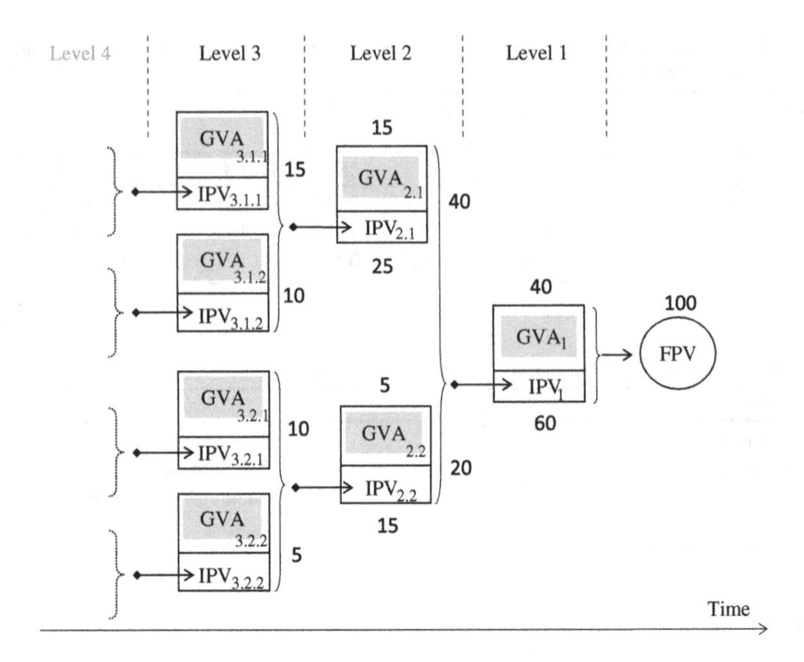

Fig. 5.3 Value chain of a hypothetical firm with the output FPV = 100

Another way of describing the value chain is depicted in Fig. 5.4, where, in the last firm, the FPV = 100, the GVA = 40, and IPV = 60. The IPV = 60 of the last firm is the sum of the GVA with the IPV of all firms of the second level, which values are 20 and 40, respectively, and so forth.

These representations of the value chain show clearly that the FPV of a firm is the sum, up to infinity, of all the GVA of the firms behind, along its value chain. As such the value of a product is the sum of the GVA of all products belonging to its value chain.

As the KTC method allows assessing the technological part of the GVA, which is T, the technological content of a product (TCP) is the sum of all the T contributions along the value chain. Making explicit the values of T along the value chain, as shown in Fig. 5.5, the TCP is given by the infinite series Eq. 5.1

$$TCP = T_1 + T_{2.1} + T_{2.2} + T_{3.1.1} + T_{3.1.2} + T_{3.1.3} + \dots \quad (5.1)$$

The output of a real firm was examined along these lines [4] scrutinizing its value chain. The firm's FPV was 100 % and its GVA (level 1) was 29 %, such that its IPV = 71 %. The GVA has given the following indexes: KI = 0.61, TI = 0.28, and CI = 0.11. At level 2, the two firms with the largest contributions (28 + 2 = 30 %) were scrutinized and their indexes evaluated. The rest of level 2 firms (41 %) was classified according to the average indexes of the economic sectors they belonged to (divisions 22, 31, 32, 33, 40, 64, 72, and 74) (8 %), and to imports (33 %). At level 3, behind the two firms identified of level 2, one firm was identified and scrutinized, the other level 3 firms being given the indexes of

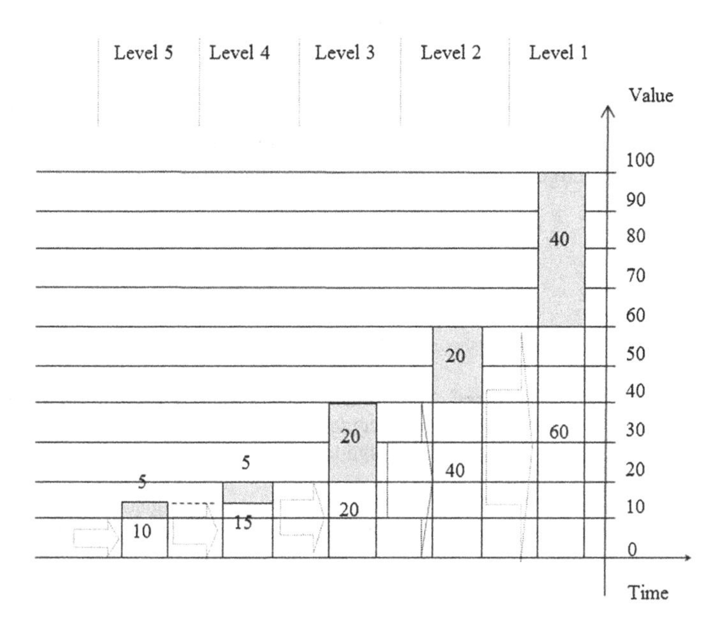

Fig. 5.4 Value chain the output FPV = 100, where the *gray shade* represents GVA and the *white* represents IPV

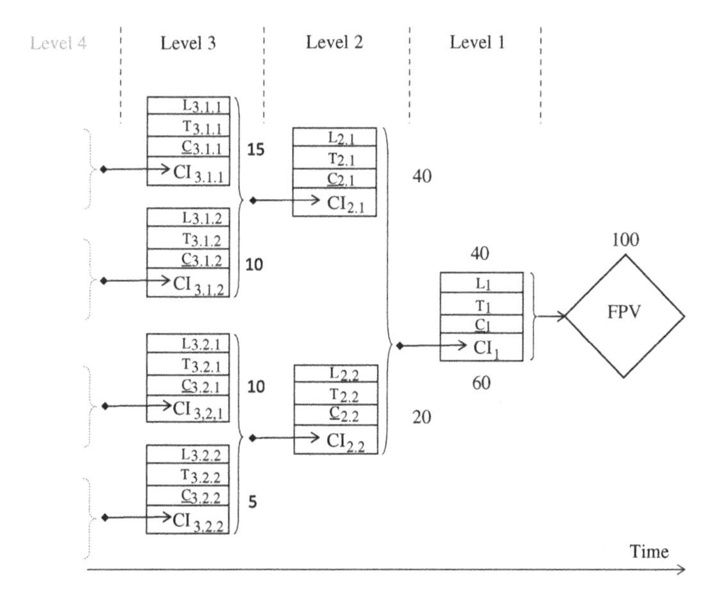

Fig. 5.5 Value chain of a hypothetical firm with the output FPV = 100. GVA is presented as the sum of $L + T + \underline{C}$

the economic sectors they belonged to. By the end of this value chain analysis, 60 % were classified rigorously and 8 % was classified according to the average indexes of the economic sectors they belong to. The remaining 32 % could not be classified. This is a theoretical framework impossible to fulfill in practical terms. However, it has the merit of pointing out the difference between the TCP and the technological contribution of one firm to its output. The same reasoning may be developed for knowledge and for capital.

References

1. Fernandes ASC (2012) Assessing the technology contribution to value added. Technol Forecast Soc Chang 79:281–297
2. Fernandes ASC (2003) Technology-driven organizations? What is that? In: Proceedings of IEEE international engineering management conference—IEMC-2003, Albany, pp 14–18
3. Hatzichronoglou T (1990) Revision of the high-technology sector and product classification. STI working papers OCDE/GD(97)216
4. Loureiro FC, Marques RV, Fernandes ASC (2002) Análise duma Cadeia de Valor Acrescentado—Fornecedores de Conhecimento, Tecnologia e Capital. Actas do 4° Encontro sobre o Valor Acrescentado pela Engenharia, Escola de Engenharia da Universidade do Minho, Guimarães Portugal (27–46) ISBN 972-8692-07-02

Chapter 6
Value Representing Technology and Knowledge

Technology and capital are knowledge driven devices used by humans to improve the capacity to better adapt and prosper in the natural and social contexts. They are socially available forms of knowledge. That capacity originates in individual human knowledge, it reveals as action and work and eventually it results in produced goods. Thus, knowledge is the basic element of survival and success, as it builds on the conscience of what we are and what we can do.

How do individual and social knowledge act and impact within society? They act in many different ways, which are delimited by a metric that adjusts their impacts on society. Indeed, the social need to exchange goods and ideas requires a scale and a balance on which trade decisions and innovations are based. That scale is used accordingly to a specific concept that substantiates and quantifies every act humans perform, in agreement with their individual need to survive and grow and balanced by the general needs of other individuals and society: That scale is part of the concept of value. People attribute value to what they need and want, and to what they produce for trading. Just as human knowledge, value has subjective and objective components. I propose a justification of how and why the existing value reflects and is proportional to the available human knowledge.

The word value originates from the Latin *valore*, which is mainly linked to the idea of being strong. Its Indo-European roots are traced through the suffix *wal* in Germanic and Slavic languages as well as Latin with a connotation also related to power and government. The concept applies to the individual, where the attribute to be more or less strong is recognized, and to the social, where it matches what is more or less needed and desirable by the group. Moreover, the concept includes attributes of two different kinds, one qualitative and other quantitative. In other words value seems to be a metric, comprising a criterion (qualitative), and a scale (quantitative). We will see that, for the economic value, a reasonably objective quantification is plausible.

A. S. C. Fernandes, *The Contribution of Technology to Added Value*,
DOI: 10.1007/978-1-4471-5001-5_6, © Springer-Verlag London 2013

6.1 The Origins of Value

Need and therefore dependency are essential components of ecosystems, with or without human presence. Plants grow up in search of light and the roots spread to find water and nutrients. The needs for water and the sun's energy are variables of the survival equation, they are parameters to which we attribute value. These parameters, essential to our scale of needs, belong to the deepest layers of the value concept, as they deal with the development of the species, a subject matter that is being adaptively filtered and innovatively optimized for 3 or 4 billion years. It is a natural development, genetically imposed, illustrating a form of local and elementary determinism. This is common to all living organisms, constituting a first factor that informs and qualifies our notion of value. This is my first work hypothesis.

In the structure of primitive societies, observed as very simple social organizations, there are two notions that fundamentally contribute to survival and prosperity of both the individual and the group: The sharing and exchange of goods; and a concern for efficiency. The first notion includes issues like safety, food, tools, and cohabitation; in the second, physical strength, reproductive aptitude, personal and group skills. This simple framework describes a primitive model of value, where the dual most essential aspects of what is needed, desirable and thus valuable are represented: (1) What is more convenient to the community; (2) and what benefits the individual. As the individual and the group are inextricably linked, they cannot present a hierarchy between them. They are the two sides of the same coin. I believe that this description of a primitive society is still valid for a complex civilization. I propose that this permanent and compulsory duality provides a second factor that qualifies in essence the notion of value. This is my second and last work hypothesis. My objective is to justify a close conceptual link between value and human knowledge.

6.2 Axiology and Other Types of Value

The theory of value is deep rooted in ancient philosophy. Value was acknowledged by the proximity to the truth, which overlapped with the ultimate end of things or the first cause. The truth was equivalent to good and its absence was the opposite, evil, principles that generated the constructions of ethics and morals mostly associated to socioreligious systems, which, in turn, stand as common ground to all civilizations. The (written) law is the first product stemming from ethical and moral conduct principles.

The Greek etymon *axios* expresses the property of having value. The equivalent English word axis means a direction or the center of a spatial dimension, which illustrates the idea that societies develop around privileged directions, their attractive axis, i.e., their main values. Axiology studies the evolution of values from a philosophical point of view.

Accordingly, even a specific geometry of value was devised illustrating an evil-good axis, pointing the good upwards and evil downwards. This geometric metaphor was used often by the Greek classics [1–3] and in the Renaissance [4]. Aristotle considered that virtues were up and vices down; Plato in his Allegory of the Cave depicted the truth as with sunlight out and up contrasting with the cave's shadows. Dante drew the axis of good and evil from up in the skies where paradise was to down deep on earth where he placed hell. This axis reminds the sun's energy source from the skies, the energy source of all living things, and the death and decomposition beneath the earth.

The most important values and virtues were common to most civilizations, what was said to make part of the natural law [5–7], though not with the same priority. For instance, Confucianism said that filial love was the first virtue, whereas for the Western culture of the same period this virtue was positioned in fourth place after reverence to gods, the spirit of the deaths, and the spirit of the nation.

Along the times, Rome, Christianity and Scholastics, the migrations of Goths from the East and the Vikings from the North did not change the fundamental values of the European older civilization. Indeed, the great majority of the population lived in the rural world where the basic agricultural production and commerce were gradually developing, keeping the vast inertia of survival, family related and local basic values. Newcomers settled down resulting in renewed societies with slow changing sets of values.

Humanism started a slow transfer from a god's centered value system to the individual capacity to build their own knowledge base and to a civil society founded on less transcendent principles, like wealth, work, competencies, and other earthly powers. Slowly, new values emerged related to work and skills triggered by new commerce opportunities. Rationalism, Empiricism, and the Kant's synthesis in his Critics definitely relocated the knowledge's source to humanity, both the individual and the social. Individuality and social values emerged as the central duality, opposition and complement, onto which nations would build their social systems and prosper. Simultaneously, industrial development and international trade became increasingly important, and with them the economy of offer and demand, a knowledge area where the concept of value developed a particular meaning.

6.3 Economic Value

Savings and wealth accumulation are as old as the history of man. Goods have to be saved, on the one hand to ensure future consumption, and on the other hand for barter with other communities. Wealth accumulation improves surviving chances and therefore it is termed as valuable. The idea of value is strongly linked to usefulness, either for consumption or for trade. This fundamental characteristic continues today to be a quality criterion of economic value. Thus wealth stands for goods with recognized value.

Considering the above, the methodological framework with which the economic value concept evolution will be analyzed here assumes value as a metric, containing a criterion and a scale, the former related to subjective aspects and the latter defining its assessment objectively.

Early forms of wealth were cattle, land, slaves, and precious metals. The introduction of the coin favored long distance trade, contributing decisively to the formation of empires where ships, armies, and strategies became visible and important forms of accumulated wealth. In the Ancient Greece, Xenophon [8, 9] explained how use value of a flute was different from its exchange value and laid down various fundamentals of economics, like showing that there cannot be wealth (value) without knowledge and also that knowledge values nothing without work. Use value and exchange value are still today two fundamental distinctions of the value concept. Aristotle [1, 10] showed how property degraded with time and the need to keep on investing to maintain its value. This is another fundamental characteristic of value: By default, it naturally degrades with time, in whatever form it is embedded. It is just like a boat rowing against the current, when stop rowing it halts and moves backwards. Similarly, humans know that the same happens with their own knowledge and with the outputs of their production.

Assessing and quantifying value using an objective scale has always been a matter of discussion. Utility and demand, on the one hand, offer and associated costs, on the other hand, stand as this matter's main sides. Thomas Aquinas, circa 1224–1274, clarified that neither the seller nor the buyer must prevail and that therefore it must be found a fair price [11]. But Duns Scotus, circa 1264–1308, felt that the fair price was an intrinsic value, which corresponded to the cost of production, wages, and other costs [12, bibliographic notes, Chap. 3]. The two scholars showed two different sensitivities to economic value,[1] the first subjective, the second objectively linked to work and other production costs. In England, Francis Bacon (1561–1626) theorized on the mercantilist capitalism, where the fair value concept evolved to the concept of market value, which was equivalent to the older exchange value [14]. Along the same line, William Petty, circa 1623–1687, defended international capital markets, stating that trade was the source of wealth creation. However, Petty proposed that value should be measured by labor and land, and land assessed by the amount of work it could provide [15, Chap. 4, no. 18–20]. He added that the market price statistically revolved around the natural value, a new concept, which would be intrinsically linked to work and to the minimum subsistence level of workers and families. This link would be developed by Karl Marx, two centuries later, and is the foundation of his theory of value objectively based on labor value. The origin of economic value was also traced to the land and labor in France by Richard Cantillon stating that "…intrinsic value of a thing in general is the measure of the land and labour which enter into its

[1] The concept of fair value is still in use today (see Directive 2003/51/CE from the European Parliament and the Council, 18 June 2003). In a similar fashion, the fair salary is defined by the Catholic Church social doctrine (Catéchisme… [13], 2434).

production…" [16, part I, Chap. 10]. At that time, Physiocracy, slowly replacing Mercantilism, became the main economic theory, seeing the land as the source of all economic value, accrediting no relevance to manufacturing and commerce. Its leading representative, Francois Quesnay [17], developed the first economic model where the new concept of added value, still today in the center of value accounting, was introduced. To the value of the land, which was a *don gratuit*, labor and capital values were added matching the final value of products.

Adam Smith returned to work (labor) as the origin of wealth creation and therefore of value. Value originated not only on wages, but on land rents and capital costs, including capital depreciation, interest, and profits. However, the part of the price that was not originated on wages would be equivalent to the effort the buyer would save to acquire that product [18, Book I, Chap. V]. He also made clear the difference between natural value and market value, the former being equivalent to total production costs, and the latter resulting from the market forces of offer and demand (Chap. VII). Finally, he endorsed the old concepts of use value and exchange value or market value (Chap. IV). Ricardo did not add anything substantial to the concept of value [19, Chap. I, Sect. II], either origins or variations, but Malthus and especially Say [20, Book II, Chap. I] moved slowly the emphasis of the origin of value from the offer side, production costs, to the demand side, and utility. Say also emphasized the importance of knowledge and entrepreneurship, prior to labor (Chap. VI), in order to create value and wealth.

By around 1871, Marginalism moved definitely the emphasis of the origin of value from the production side to the utility sensed on the demand side, though without overlooking the importance and relevance of production costs. Carl Menger and Friedrich von Wieser (1893) [21], from Austria, Jevons [22], from England, and Walras [23], from France, laid the foundations of this new theory, beginning the neo-classic economic era. The degree of the user's potential satisfaction, the utility, would be the main criterion of a valuable good. This new vision carried a higher weight of subjectivity. However, Walras subsequently developed a general equilibrium theory, defining effective demand and effective offer, the latter coinciding with the utility curve. The two curves were represented by two equations that lead to a mathematical result, which is the transaction value. A rigorous mathematical theory of value would need another 50 years to be complete [24]. This corresponds to a suitable scale to assess and quantify the value of a product. The price is the transaction value per unit of the transacted goods. Alfred Marshall in Cambridge developed and analyzed extensively Walras markets' general equilibrium, what would be known as local equilibrium theories, explaining how both production costs and utility contribute to find the transaction value, just as the upper and under blades of a pair of scissors contribute to cut a piece of paper [25, Book V, Chap. 3, point 7]. The Marshallian cross representing the two curves effective demand and effective offer became the transaction value paradigm up to the present days.

It can be concluded that if considering value as a metric, its criteria comprises both the offer side, with its corresponding objective productions costs, and the demand side representing the consumer's subjective utility. Moreover, that metric's scale is given locally by the result, the price, found in the process of an economic transaction. Finally, whatever criteria are assumed, the economic value can only be quantified when an economic transaction takes place. At that point, once established the value of a product, one can trace how that amount of value was built, how much was objective and how much was subjective, how was it added along the value chain, and how much is consumed or substitutes and restores depreciated value or, alternatively, how much represents new created value (CV). Objective and measurable concepts of value consumed, value restored, and value created will help to understand how close to knowledge the concept of value is.

6.4 Value Consumed, Restored and Created

What is sometimes referred to as Schumpeterian economics has the concept of innovation at its heart, and the importance of innovation can only be understood within the context of the value concept. Innovation is consensually recognized as the main growth engine, and can be attained by a service or a product that brings extra value to society. What Schumpeter [26, Chap. VII] pointed out was that new processes and products, besides incessantly revolutionizing the economy structure, they also incessantly destroy the old one. How can that extra value delivered to society be measured? Methods like net present value (NPV) do it easily, comparing investment and returns when using or not using the innovative good, or comparing returns to those of a reference investment. This is typically linked to local management decisions on an investments' portfolio. But in a larger scale how can one objectively know if a firm, a sector or a whole economy has been innovating? I would answer this question by computing the value created by that firm, sector, or economy. The problem with this simple answer is that the concept of value creation is not consensual, and thus there is no standard accounting algorithm to compute it.

I start by clarifying the concept of value added,[2] and proceed explaining what I consider as value consumed, restored, and finally created value (CV). These notions will provide the rational to understand how creating or destroying value within society reflects on knowledge and vice versa.

This rational builds on the following assumptions:

1. A society produces and consumes goods and services to which value is attributed.
2. Value cycles may be described using flows and stocks of value.
3. Consumption sustains and develops human knowledge.

[2] By value added it will be used the economic concept of gross value added (GVA) as defined in ESA [27], 1.15 and 8.89.

4. The society's stock of value has two components: assets net value and human knowledge value.
5. All forms of value depreciate with time.

6.4.1 Production, Consumption and Investment

Along the value chain, each economic unit adds value to the intermediate products to build the final products value, as shown in Fig. 3.2. The final products value may be consumed by the final client or may act as an intermediate product value that will feed the next economic unit along the chain. In a whole economy, the added value by all economic units makes the economy's gross value added (GVA), and that is the produced value by that economy.

Simplifying, without restricting the model's validity, we can model an economy with no state and without exchanges with other economies. In this economy, the produced value equals the families' income value and that income value equals their expenditure value.

This is depicted on Fig. 6.1, where arrows represent value flows, out of and back in the stock of value. Following Fernandes [28], the resources value is the stock of value, where one may distinguish two parts: One is the capital's value, which is quantified on all the economic unit's balance sheets; and the other is the families' knowledge value, which is unquantifiable. Production uses two flows

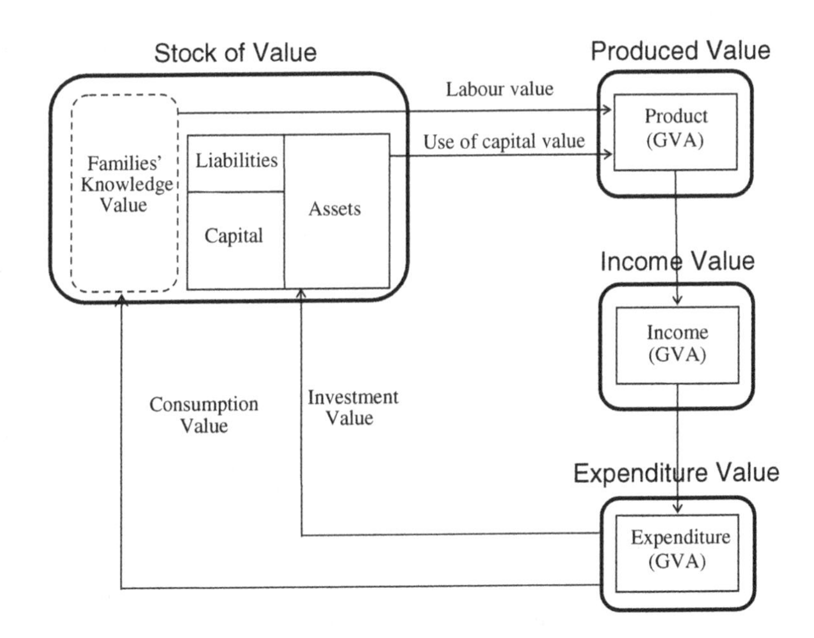

Fig. 6.1 Value cycle (Fernandes [28], reproduced with permission from Inderscience Publishers)

of value: Labor value, which one might think of as originated in the families' knowledge stock and materialized as action and work; and the value corresponding to the use of capital, originating in the capital stock and materializing in its use. As capital stock, I consider the capital value computed as the net assets value minus liabilities. Both the labor value and the use of capital value are objectively accountable and so is the produced value (or the product, or the GVA). As said above, the produced value is also the families' income value and their expenditure value. The families' expenditure has two parts: Consumption and savings (investment). Consumption value corresponds to what families expend in order to sustain and develop their knowledge, here including their physical existence. Investment is the value of capital forms that will more or less compensate the capital value depreciated along the previous cycle. The value cycle starts with value outflow from the stock into production, income and expenditure, and ends with value inflow back to the stock. Macroeconomic national accounting standards quantify precisely all the described flows and the part of the stock related to assets [27].

Where is the value consumed, restored, and created? The value consumed is the value related to consumption and can be easily quantified. For instance, in Portugal between 1997 and 2003, final consumption was about 95 % of the GVA [29]. There are no direct ways of calculating the amount of consumption needed to keep stable the families' stock of knowledge or to increase it. Indirect measures assess health and education metrics and similar issues. On the other hand, we know precisely how much capital value is used in a cycle, and thus how much value did flow out of the stock into production. In other words, we know how much investment value we need in order to keep constant the capital stock of value. Hence, the value restored is the consumption value plus part or the total of the investment value, which can be precisely computed. Finally, where in the cycle is value created or destroyed and how can we calculate it?

6.4.2 Minimum Value to Return and Created Value

To help defining CV and establishing an algorithm to quantify it, I will start by introducing the idea of the minimum value to return (MVR). This is here defined as the quantity of value to be returned to the stock such that, along one cycle, the stock retains the capacity to regenerate the same present value on the next cycle. One might think, at this point, that to be able to regenerate the same present value on the next cycle means to have the same value at the end of the cycle as there was at the beginning; but no, because that would be the case of a context with no competition and no inflation. What it means is that the stock of value must keep its present value, like in the NPV algorithm that was referred to above. Thus, a constant present value implies that there must be a value returned that equals the value expended plus an interest. Accordingly, this definition of MVR implies that a minimum interest rate should be established in order to act as a reference. It will be seen below what this interest rate might be.

Finally, we are able to define and quantify the CV. It is the difference between the expended value (which equals the GVA) and the MVR. When GVA is higher than the MVR there is value created; if it is smaller there is value destroyed.

As pointed out, there is one uncertain constituent in this cycle, which is the families' knowledge value. As this value is not measurable, accounting systems do not reveal its balance. It will be seen that this is not impeditive for computing both the MVR and the CV, even if it introduces some ambiguity. Actually, it is precisely within the families' knowledge stock where the value creation likelihood is triggered. How does that happen? It happens by what it is known as innovation. This happens when human knowledge devises a process (incremental or new) or a product that originates value creation. Figure 6.2 describes this situation, where innovation takes place and there is value created. In this cycle, there is an initial stock of value and there is a final stock with a higher value, as the GVA exceeds the MVR and so there is a positive CV.

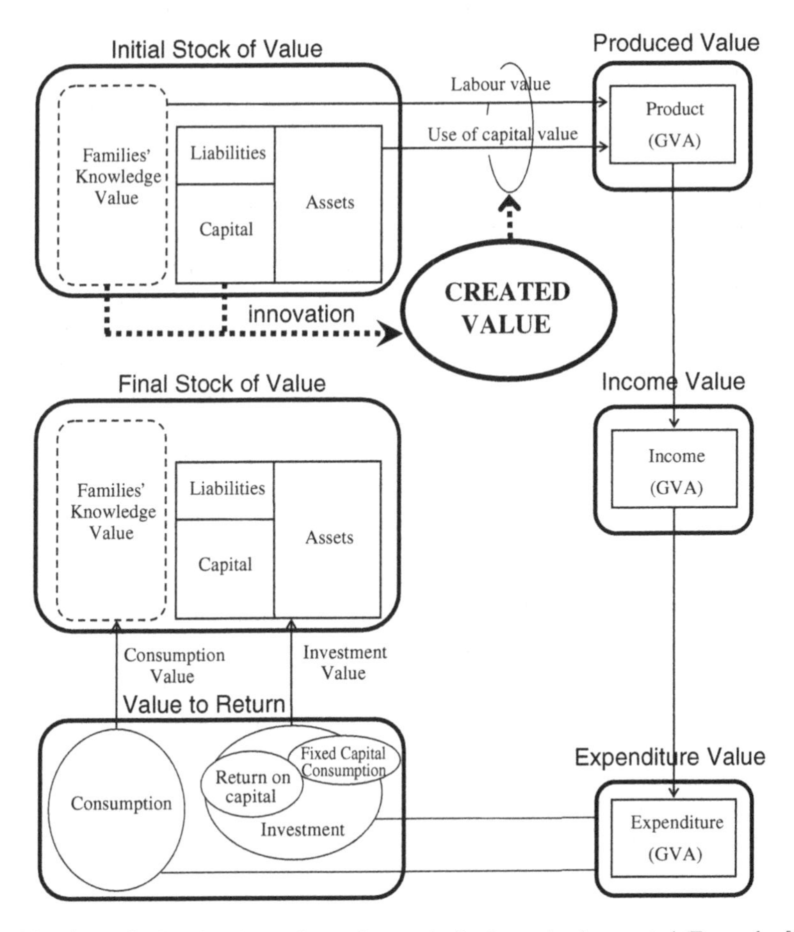

Fig. 6.2 Flows of value showing an increasing stock of value and value created (Fernandes [28], reproduced with permission from Inderscience Publishers)

Using the accounts as in Chap. 3, both the GVA and the MVR can be written as sums of accounts and easily computed for firms, sectors, or whole economies:

$$\mathrm{GVA} = 6 + 7 + Y + 21 - [(16 + 9\,\mathrm{to}\,11) - (12 + 13 + 17)] \qquad (6.1)$$

$$\mathrm{MVR} = 6 + 7 + 13 + Y + r \cdot C \qquad (6.2)$$

Equation 6.1 reads as follow: GVA equals the sum of (6) wages and social security costs, plus (7) depreciation on fixed assets and provisions, plus (13) interests and other charges on financial debts, plus (*Y*) taxes on profits, plus (21) profit for the financial year, minus [(16 + 9/11) – (12 + 17)]. This last term is incomes minus costs of financial and extraordinary activities; and so it is the profit of financial and extraordinary activities.

Equation 6.2 reads as follows: The MVR equals the sum of (6) wages and social security costs, plus (7) depreciation on fixed assets and provisions, plus (13) interests and other charges on financial debts, plus (*Y*) taxes on profits, plus a new term (*r.C*) that reflects the minimum return on the net capital used in the economic process under evaluation. *C* is the capital net value and *r* is the return on capital coefficient (reference interest rate). This last parameter has to be valued according to the local situation, like inflation and objective investment opportunities. The last term (*r.C*) is the minimum return on the capital that capital owners invested. How much should this return be? This question can have a very (apparently) complete and sophisticated answer or it may be simplified to an uncomplicated working number. The first approach would consider *i* different types of capital, like, cash, technologies, buildings and so on, and consequently return coefficients r_i specifically calculated for each one. Also, for technological and other capital forms, either the book value or the market value could be considered. And again, the market value will have to take into consideration the specific investment opportunity to which those assets are committed to and the corresponding discounted cash flows (DCF). This last calculation will depend on a number of market parameters. This approach becomes very easily dependent on tricky and subjective forecasts. The second approach, which I favor here, considers for *r* (interest rate) an average national number for each year and, for *C* (capital and reserves), its book net value. For the coefficient r, it is proposed the year's average EURIBOR[3] interest rate at 12 months (or an equivalent interbank rate in other economic regions). In this way, an approach that is both objective and simple is favored such that the MVR is easily calculated. Finally, the CV may be computed:

$$\mathrm{CV} = \mathrm{GVA} - \mathrm{MVR} = \{21 - [(16 + 9/11) - (12 + 17)]\} - (r \cdot C) \qquad (6.3)$$

Following Leal and Fernandes [30], we may analyze the value created in some sectors of a few European countries and illustrate this new concept. It was considered a set of nine sectors covering the activities related to Mining and Quarrying

[3] Euribor® (Euro Interbank Offered Rate) is the rate at which euro interbank term deposits within the euro zone are offered by one prime bank to another prime bank. http://www.euribor.org/default.htm (accessed May 2008).

(sector B), Manufacturing (C), Electricity (D), Water (E), Construction (F), Trade (G), Transport (H), Hotels and Restaurants (I), and Communications (J). For the European Union countries, data were used from the Bank for the Accounts of Companies Harmonized (BACH) [31]. The data collected included information relating to Austria, Germany, Belgium, Spain, France, The Netherlands, Italy, Poland, and Portugal. It was adopted the year's average Euro Interbank offered rate (EURIBOR) for the period of 12 months as the minimum return on capital coefficient r.

Computing the ratio CV/GVA, results are depicted in Fig. 6.3. We can see that Austria and Poland have had the highest average percentages of CV. On the other hand, Belgium has cycles of value destruction for most of the period analyzed, reaching only in 2004 and 2005 stages where no value is created or destroyed. In almost half of the period considered, the countries with the worst performances are Italy, from 2004 to 2005, and Portugal, from 2005 to 2008. Italy did not register any cycle with positive CV. It is relevant to highlight that on 2008 all countries show a sharp drop and between 2002 and 2004, a generalized improvement was registered, that lead all countries, except Italy, to positive amounts of CV.

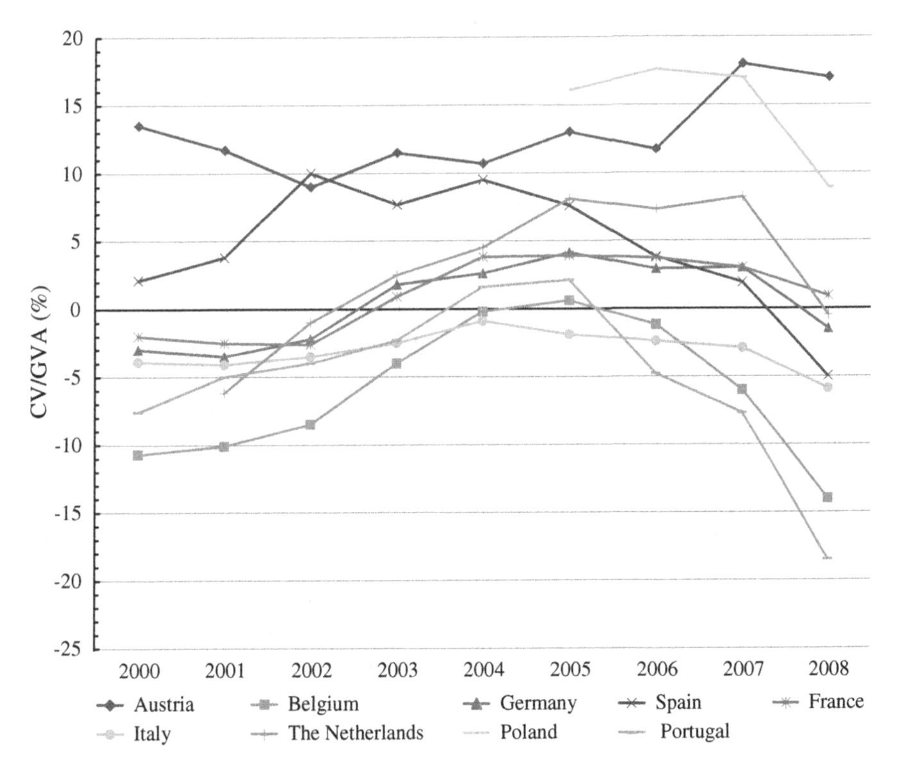

Fig. 6.3 Ratio created value/gross value added for nine EU countries—results group together sectors B, C, D, E, F, G, H, I, and J (NACE rev. 2)

Concluding, it was presented and analyzed the concept of (economic) value, in its main expressions, what is relevant to understand how value represents knowledge:

- Stock of value: assets net value and human knowledge value.
- Flows of value: activities that transfer value.
- Produced value: value (GVA) that flows from stocks to goods and services, through the use of labor and capital.
- Income value: value received by the families both from their work and as a return from their capital.
- Expenditure value: value expended by the families for consumption and investment in capital.
- Restored value: the consumption value plus the fixed capital consumption value.
- MVR: the quantity of value to be returned to the stock such that, along one cycle, the stock retains the capacity to regenerate the same present value on the next cycle.
- CV: the positive difference between the GVA and the MVR.

6.5 The Cycle Knowledge-Value-Knowledge

In a comprehensive and socially meaningful economic analysis, human society must be considered as the beginning and the end of the economic system. Therefore, any model to represent flows and stocks of value must reflect people's gains and losses.

Human knowledge is the concept that better aggregates all dimensions of the humans and Nature interaction, and each dimension works for the purpose of surviving and prosper. As such, human knowledge, with Nature in the background, triggers the appropriate actions and work, both individually and within their complex social system, in order to achieve those ends. The social fabric, the economic system, the culture, as well as particular parts like capital and technological forms are instruments and ways of assisting the path of anticipation for prosperity. Capital and technology forms are human knowledge embedded and encrypted in natural materials, serving the purpose of adapting to a special situations and multiplying the efficiency of humans' work.

The simplest cycle shows that humans act and work, and subsequently consume. As such, human knowledge triggers work and in return is supported and develops through consumption. This is depicted in Fig. 6.4, which shows the cycle of knowledge, beginning and ending in humans: Knowledge–work–work products–consumption–(renewed and more) knowledge. In our present culture, the cycle described: Work–work products–consumption is assessed objectively using the concept of value. Thus, the conclusion that value represents human knowledge along the cycle and so is directly and univocally comparable to it.

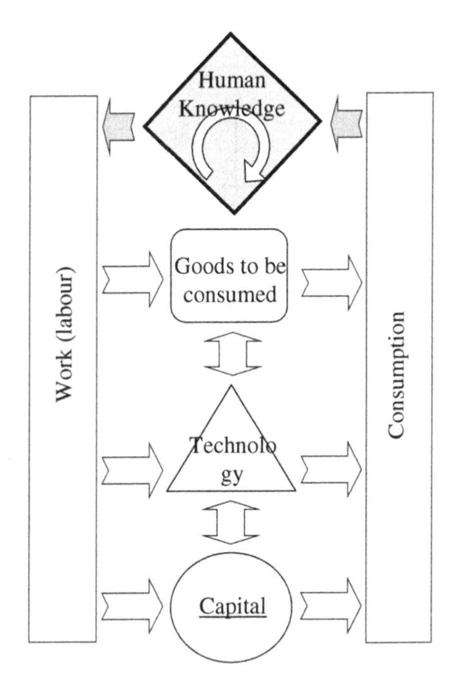

Fig. 6.4 The cycle knowledge-value-knowledge (adapted from Fernandes [28], reproduced with permission from Inderscience Publishers

6.5.1 How Value Represents Knowledge

A more objective reasoning can be proposed, which will prove that value is equivalent to knowledge, assuming that labor value is the best possible proxy to assess human knowledge value and so human knowledge itself.

Let us take the final product value (FPV) = 100 of a specific good representing all possible final consumption goods. The concept of value chain, as presented in Figs. 5.3 and 5.4, will also be used. The value chain of a product to be consumed represented in Fig. 5.3 can be drawn in a slightly different way, described in Fig. 6.5.

The GVA was replaced by its two main components: Labor value L and a surplus SP, as shown in (6.4):

$$\mathrm{GVA} = L + \mathrm{SP} \tag{6.4}$$

As such, the FPV = 100 equals the GVA = 40 plus the IPV = 60 of the second level, and the latter equals the FPVs of the third level firms, and so on. The FPV of the specific good under analysis can then be described as shown in (6.5):

$$\mathrm{FPV} = L + \mathrm{SP} + \mathrm{IPV} \tag{6.5}$$

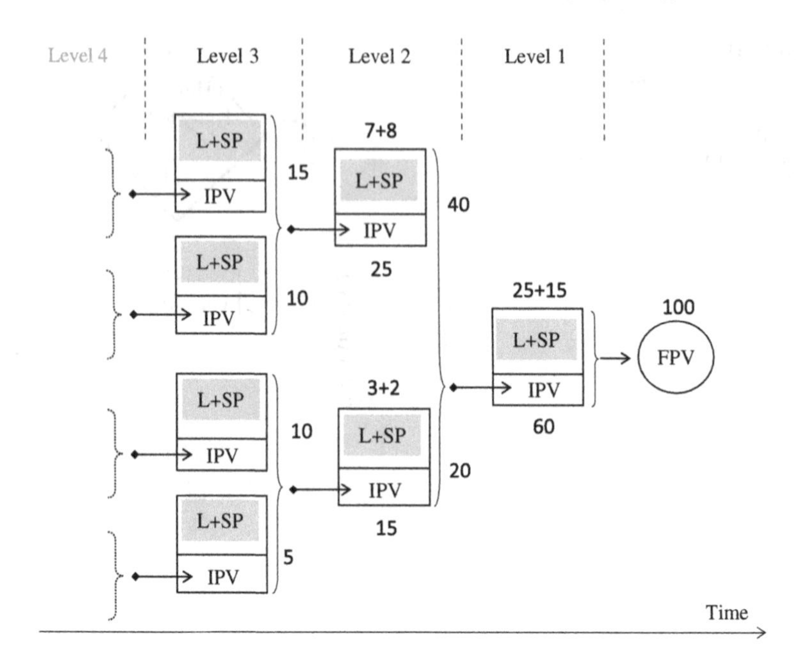

Fig. 6.5 Value chain of a final consumption good, which final product value is FPV = 100

It was proved in Chap. 5 that the FPV is the sum of the GVA of all products belonging to its value chain. In fact, the intermediate product values (IPV) are the values of intermediate products, which are goods that can be subject to exactly the same value chain analysis. Hence, IPV can also be described as a sum of terms L and SP and other IPVs. Mathematically, the expression (6.5) can then be described as (6.6), where the rest is as small as we want, depending on how many levels we use to analyze the value chain. Assuming that the value chain has infinite levels, if we would tend to take those infinite number of levels the rest would tend to zero.

$$\mathrm{FPV} = \sum L + \sum \mathrm{SP} + \mathrm{rest} \tag{6.6}$$

This conclusion also expresses what was proven in Chap. 5, that the value of a product FPV is the sum of the GVA of all products belonging to its value chain.

Now, let us consider that the surplus SP is proportional to capital C, such that $\mathrm{SP} = m \cdot C$. The value of capital C being the value of goods (assets) that performs the task of capital, so C is a value that can also be analyzed in the same way. The factor m is a parameter dimensionless and almost always positive [32]. Thus, (6.6) can be written as (6.7):

$$\mathrm{FPV} = \sum L + \sum m \cdot C + \mathrm{rest} \tag{6.7}$$

Considering C as a fictitious one single asset contributing the fraction $m_i \cdot C$ in each part i of the chain, along the whole value chain (and so totally used), the sum

$\Sigma\, m_i \cdot C$ will necessarily equal to C. In other words, $\Sigma\, m_i = 1$. Thus (6.7) is written as (6.8):

$$\text{FPV} = \sum L + C + \text{rest} \tag{6.8}$$

This asset's value C is the value of a good and so it can be analyzed in the same way, hence C itself may also be expressed by a similar identity, then with another C_j, which would value less than C, and another L_j. And so on, repeating the same logic. Considering an infinite number of such steps, we would end up with the following expression (6.9):

$$\text{FPV} = \sum L + \sum L_i + \sum L_j + \sum L_k + \cdots + \text{rest} \tag{6.9}$$

As such, with an infinite value chain, the rest = 0 and FPV equals the whole amount of labor value summed up along the whole value chain.

This reasoning implies a level of abstraction that is sometimes difficult to follow. To put it clearer, I propose a simple example describing a value chain of the activity of a small society of fishermen. It is a closed extremely simple economy without state producing along 1 week, from Monday to Saturday, and resting on Sunday. On Monday, they start the week's work with one new fishing net, which value is C. The net has a working life of 5 days, such that each fishing day, Monday to Friday, it wares out one-fifth of its value. The fishermen work is fishing for their daily consumption and for consumption on Saturday and Sunday. On Saturday, the fishermen do not go fishing; instead they stay at home producing a new fishing net, for what they only need their knowledge and work plus woof made out of bark from a few local trees.

According to the main economic identities, the GVA produced every day is shown in Table 6.1, bottom line.

The GVA produced from Monday to Friday is calculated in (6.10). This is the value of the fish produced and consumed. As there are no IPV, this GVA = FPV.

$$\text{GVA} = \text{FPV} = \sum L_{2\,\text{to}\,6} + \sum m_{2\,\text{to}\,6} \cdot C = \sum L_{2\,\text{to}\,6} + C \tag{6.10}$$

Moreover, C is produced on Saturday using the labor value L_{Sa}, such that C values L_{Sa}. Thus

$$\text{GVA} = \text{FPV} = \sum L_{2\,\text{to}\,6} + L_{\text{Su}} \tag{6.11}$$

Table 6.1 Day's GVA produced

C working life is 5 days						
Monday	Tuesday	Wednesday	Thursday	Friday	Saturday	Sunday
L_2	L_3	L_4	L_5	L_6	L_{Sa}	–
$m_2 \cdot C$	$m_3 \cdot C$	$m_4 \cdot C$	$m_5 \cdot C$	$m_6 \cdot C$	–	–
$L_2 + m_2 \cdot C$	$L_3 + m_3 \cdot C$	$L_4 + m_4 \cdot C$	$L_5 + m_5 \cdot C$	$L_6 + m_6 \cdot C$	L_{Sa}	–

This is the same result as showed in (6.9). What was proved is that the value of final consumption products is the value of the labor used along their value chain. Then and finally, assuming that labor value is the best possible proxy to assess human knowledge, we may take the conclusion that value is the metric that societies use to assess knowledge.

6.5.2 Value as the Criterion for Knowledge

Economic value was concluded to be a metric. In Sect. 6.3, it was said that the metric had criteria and a scale, the former being the user needs from the demand side and the producer needs from the offer side; and the latter, the scale, was set continuously in every transaction, locally or globally, by its corresponding price.

If economic value assesses knowledge that is economically meaningful, as concluded above, then value, in general, might assess human knowledge, not only from the restricted economic point of view, but also from a socially perspective, as pointed out at the first two sections of this chapter. Even if the above demonstrations are not complete, as it is always the case in social sciences, I am inclined to believe that value is the basic metric for assessing human knowledge, and thus value reflects the available human knowledge, that is, the knowledge embedded in technology and capital and the knowledge to act and work in order to producing goods to be consumed. Being true, that would bring economics to the center of social evolution, the Schumpeter vision.

References

1. Aristóteles (1999) Nicomachean Ethic. Batoche Books, Kitchenner (trans: Ross WD). http://socserv2.socsci.mcmaster.ca/%7eecon/ugcm/3ll3/aristotle/Ethics.pdf. Accessed Sept 2012. Also in http://classics.mit.edu/Aristotle/nicomachaen.html (Book IV). Accessed Sept 2012
2. Aristotle (1952) Virtues and vices. Aristotle in 23 volumes vol 20 (trans: Rackham H). Harvard University Press, Cambridge. OCLC: 3906945. http://www.perseus.tufts.edu/hopper/text?doc=Perseus%3Atext%3A1999.01.0062&redirect=true. Accessed Sept 2012
3. Plato (2001) A República. Lisboa: Edition by Fundação Calouste Gulbenkian. Translated from Greek by J. Burnet, Platonis Opera T. IV, Oxonii e typographeo Clarendoniano, 1949, by Maria Helena da Rocha Pereira
4. Alighieri D (1995) A Divina Comédia. Lisboa: Bertrand Editora, Lda. Translated by Vasco Graça Moura (Original from 1310–14). Also in http://www.italianstudies.org/comedy/index.htm (trans: James Finn Cotter). Accessed Sept 2012
5. Cicero (52 BC) De Legibus (On the laws). http://www.fordham.edu/halsall/ancient/cicero-laws1.asp. Accessed Sept 2012
6. Cicero (51 BC) De Republica (The republic). http://www.fordham.edu/halsall/ancient/cicero-republic1.asp. Accessed Sept 2012
7. Grotius H (1583–1645) The law of nature. http://www.humanistictexts.org/grotius.htm. Accessed Sept 2012

8. Xenophon (1979) Xenophon in seven volumes. Harvard University Press, Cambridge. Works on socrates, symposium, Chaps. 3 and 4. http://www.perseus.tufts.edu/hopper/text?doc=Perseus%3Atext%3A1999.01.0212%3Atext%3DSym.%3Achapter%3D3%3Asection%3D5. Accessed June 2012
9. Xenophon (1979) Xenophon in seven volumes. Harvard University Press, Cambridge. Works on socrates, economics, Chap. 1 http://www.perseus.tufts.edu/hopper/text?doc=Perseus%3Atext%3A1999.01.0212%3Atext%3DSym.%3Achapter%3D3%3Asection%3D5. Accessed June 2012
10. Aristotle (1935) Aristotle in 23 volumes vol 18 (trans: Armstrong GC). Harvard University Press, Cambridge. http://www.perseus.tufts.edu/cgibin/ptext?lookup=Aristot.+Econ.+1344b. Accessed Sept 2012
11. Aquinas ST (1947) The Summa Theologica. In: Benziger Bros (ed) (trans: Fathers of the English Dominican Province). Question 77 in http://www.ccel.org/a/aquinas/summa/SS/SS077.html. Accessed Mar 2012
12. Spiegel HW (1999) The growth of economic thought. Duke University Press, Durham & London, 3ª ed
13. Catéchisme de L'Église Catholique (1998) Paris: Centurion/cerfs/Fleurus-Mame
14. Bacon F (1625) The essays or counsels, civil and moral l-Essay—XLI-On usury. http://socserv.mcmaster.ca/econ/ugcm/3ll3/bacon/index.html. Accessed Sept 2012
15. Petty W (1662) A treatise of taxes & contributions. London, http://socserv2.socsci.mcmaster.ca/~econ/ugcm/3ll3/petty/taxes.txt. Accessed Sept 2012
16. Cantillon R (1755) An essay on commerce in genera. http://socserv.mcmaster.ca/econ/ugcm/3ll3/cantillon/index.html. Accessed Sept 2012
17. Quesnay F (1985) Quadro Económico. Lisboa: Fundação Calouste Gulbenkian. Published for the 1st time in 1758. Also in http://socserv.mcmaster.ca/econ/ugcm/3ll3/quesnay/tabeco.htm. Accessed Sept 2012
18. Smith A (1956) Inquiry into the nature and causes of the wealth of nations. Collier & Son Corporation, New York (first published in 1776)
19. Ricardo D (1821 3ª edition) On the principles of political economy and taxation. Batoche Books, Kitchener, Ontário, 2001. http://socserv2.socsci.mcmaster.ca/%7eecon/ugcm/3ll3/ricardo/Principles.pdf. Accessed Sept 2012
20. Say JB (1841) A treatise on political economy; or the production, distribution, and consumption of wealth, 1st edn 1803. Batoche Books, 52 Eby Street South Kitchener, Ontario, Canada. Translation from French 4ª edn in http://socserv2.socsci.mcmaster.ca/%7eecon/ugcm/3ll3/say/treatise.pdf. Accessed Sept 2012
21. Wieser F (1893) Natural value. Translation from German edition of 1889 by Christian A. Malloch. http://socserv2.socsci.mcmaster.ca/%7eecon/ugcm/3ll3/wieser/natural/index.html. Accessed Sept 2012
22. Jevons WS (1866) Brief account of a general mathematical theory of political economy. J Roy Stat Soc Lond. http://www.marxists.org/reference/subject/economics/jevons/mathem.htm. Accessed Sept 2012
23. Walras L (1883) Théorie mathématique de la richesse social. http://www.intertic.org/Classics/bertrand.pdf. Accessed Sept 2012
24. Debreu G (1959) Theory of value. Cowles foundation monograph. Yale University Press, USA
25. Marshall A (1890) The principles of economics. http://socserv2.socsci.mcmaster.ca/%7eecon/ugcm/3ll3/marshall/prin/index.html. Accessed July 2012
26. Schumpeter J (1976) Capitalism, socialism and democracy. Routledge, London
27. ESA 1995 (1996) European system of accounts. Office for official publications of the European Communities, Luxembourg
28. Fernandes ASC (2010) (Biblical) Creation of value. Int J Eng Manage Econ 1(1):46–67
29. Fernandes ASC (2007) Mais Conhecimento e Tecnologia para Desenvolver a Economia Portuguesa. Fundação Calouste Gulbenkian, Lisboa
30. Leal V, Fernandes ASC (2011) Value creation in Portugal. In: Proceedings of the IEEE Eurocon 2011 International conference, 27–29 April, Lisbon, Portugal

31. Banque de France & European Committee of Central Balance-Sheet Data Offices (2010) BACH and BACH-ESD: user guide June 2010. http://www.bachesd.banque-france.fr/?page=documentation. Accessed Sept 2012
32. Fernandes ASC (2009) The management impact on value added—an approximate measure. In: Proceedings of the IEEM 2009—IEEE international conference on industrial engineering and engineering management, Hong-Kong (1391–95)

Chapter 7
Key Conclusions

The stated main goal was to contribute to better understand the meaning of technology and to offer one economic model where its role could be operationalized. In that way, it would be possible to objectively compute the contribution of technology to added value and to economic growth.

The two main issues that had to be resolved were pointed out as, first, a semantic misunderstanding and lack of consensus about the meaning of the word technology, and second a dimensional problem and consequent lack of trust in many production functions and total factor productivity analyses, a problem that contributed to an added misperception and ambiguity of the technology role in a production process.

It was shown that many analyses using the Cobb-Douglas production function, namely the Sollow model, result in misleading conclusions when comparing outcomes. This is due to different dimensions of the parameter A, which is associated to technical progress or technology, when used in different economies and different periods of time (α is different for different economies and for different years). Both exogenous and endogenous technology growth models reveal the same problem as well as many different types of regression analyses. For identical reasons, total factor productivity analyses produce misleading results, the exception being the multifactor productivity analysis performed by OECD. Besides this dimensional problem, it was exposed the obvious ambiguity when interpreting the meaning of the parameter A, referred to as technology, knowledge, technology knowledge, technical progress, a residual or everything else besides labor and capital, etc. It became apparent the overlap between the concepts of technology, knowledge, and capital. One corollary exerted from this exercise was that, if we could have the three concepts as independent and autonomous, we would be able to model a value adding process by adding each one's value contribution. Accordingly, a linear model to represent that process was proposed. This was described in Chap. 2.

As the concepts of technology, knowledge, and capital were deconstructed and reconstructed as operational concepts, it was possible to model a production

A. S. C. Fernandes, *The Contribution of Technology to Added Value*,
DOI: 10.1007/978-1-4471-5001-5_7, © Springer-Verlag London 2013

process such that the three new concepts present an autonomous role, hence identifying the role of technology in value adding. The new KTC linear production function was compared with the standard accounting rule for the GVA and established the model's algorithms to compute the technology contribution for value added, as well as the contributions of knowledge and capital. However, the success of reconstructing the concepts was not thorough, especially when separating technology from capital, being the border line somehow dependent on the firm or the sector under study. In other words, the results showed that there is not an absolute separation of the two operational concepts technology and capital. For instance, there might be assets that play the role of technology in one firm and may play the role of capital in another.

In their fundamental attributes, the operational concepts knowledge and technology were discriminated as follows: (1) Human's knowledge is dynamic and exists within the human's mind; (2) Technology is a produced good where human's knowledge is embedded but in a static form.

Therefore, an organized thought, a new idea, a theory, a schematic image, a set of principles, a formulation of a question, a decision sequence, any image or other forms of memorized data, a melody, and so forth are knowledge as far as they exist in the human's mind, because they represent data and information constructed and resident there. They are knowledge.

On the other hand, a message written on a paper, a sketch, a hand-written or CD-recorded speech, a built object, a programmed decision algorithm, a technological rule, a methodological written sequence and so on, up to tools, instruments, machines, and large systems are technological forms, because they are exterior and already independent of the minds that created them; they are objective forms in a material state. They are directly available by society and are potentially understood and used by many people. They are technology.

Capital forms are objective and identifiable material goods with the same attributes as technology, but with a much larger flexibility in the role they play in a production process. To differentiate technology from capital, it is important first to remember that capital and technology attributes mostly coincide. They are both produced goods not to be consumed, with specific functions in the production process. They both have static knowledge content and often a recognizable shape. Yet, one attribute may contribute to discriminate between them: Capital has a more flexible applicability than technology. Capital's flexible applicability means that it is easily transformable (liquidity). On the contrary, a technology form has a specific function and cannot be used for anything else. Therefore, the boundary between technology and capital was traced with the following criterion: That form that is versatile in its functionality and is easily transformed is a form of capital. If its function is well defined and cannot be easily transformed, it is a technological form. As such, examples of technological forms are patents, blueprints, reports, tools, instruments, machines, systems, and all sorts of equipment, whereas capital forms are typically buildings, money, and credit. Also, land should be considered a basic form of capital, as far as it is involved in a production process. This border is not absolute, and sometimes a static border line is difficult to draw.

The KTC model allowed as well an important conclusion: Writing its production function in differential form, it becomes clear what the economic growth conditions are: A simultaneous growth of labor value and knowledge productivity value, being conceivably the two growth rates uneven, depending on socioeconomic local conditions. The optimum seems to correspond to increases in both, sometimes higher for KP, sometimes higher for L. Increases of both L and KP imply, in conclusion, an increase in investment. This was described in Chap. 3.

The new model and algorithms were developed and successfully applied to different economic sectors and economies, computing the technology index, what reflects the relative technology dependence, as well as the knowledge index and the capital index. Sectors like education, construction, computer and related activity, other business activities have especially high values of KI. The highest technology index TI is found in sectors as renting of machinery and equipment, different types of transport, electricity, telecommunications, etc. High values for the capital index CI are found in real estate activities, manufacturing of coke and refined petroleum products, electricity, etc. When comparing manufacturing sectors of different economies, surprising results were found as Germany and France showed much higher values of KI and smaller values of the technology index. We conclude that, as they have higher average wages, a higher KI implies a smaller TI and/or CI. This was described in Chap. 4.

The results of the technology index were compared to the OECD's technology intensity and it was shown how TI reflects better the technology dependence of firms, sectors or economies. A statistical analysis of the TI results also allowed establishing a metric to propose a specific taxonomy for technology dependence. The suggested levels would be: Low technology, medium–low technology, medium–high technology and high technology. This model also permitted the calculation of the technological content of a product, considering the whole value chain technology dependence, what is especially relevant for analyzing clusters. An important conclusion concerning the GVA could also be taken: The value of a product is the sum of the GVA of all products belonging to its value chain. This was described in Chap. 5.

Through an analysis of the concepts of value and economic value, we were lead to the idea that production and consumption values reflect what is needed to keep and develop society's knowledge base. Along these lines, the concept of created value was proposed and computed for a number of countries, from 2000 to 2008, clearly showing how much value is destroyed during recessions. Moreover, it could be concluded that the value of final consumption products is the value of the labor used along their value chain, what has been mentioned by several economists along the last centuries but never proved. Finally, an interpretation of the socioeconomic dynamic fabric leads to suggest a knowledge-value-knowledge cycle, which shows why value can be interpreted as a metric for knowledge. This was described in Chap. 6.

Annex

Technology, Knowledge and Capital: Building Operational Concepts

This annex shows how the methodology is applied to analyze the concepts of knowledge, technology, and capital and how the respective operational concepts can be built. It has already been explained the different roles a concept and an operational concept may perform in scientific reasoning and model building. It follows Fernandes [1, 2].

The method takes five steps, A to E:

A. To analyze and to list each concept's attributes as observed in the texts of different areas of knowledge, different authors, and along the time.
B. To find the attributes that are common among those areas and constant through time, most probably the ones that better characterize each concept.
C. Once the concepts are deconstructed in their fundamental attributes and respective extensions, to enunciate a criterion of reconstruction of the new operational concepts, accordingly to a pre-set purpose.
D. To build the new operational concepts using the fundamental attributes and the appropriate extensions, and propose a definition and an explanation for each one.
E. To clarify the criteria for distinguishing the boundaries among the three operational concepts, such that they could be objectively identified.

The criteria to be used for reconstructing the three operational concepts are set as follows:

1. To allow an objective identification, a quantification and the understanding of the roles of each one in a production process;
2. The final form of the operational concepts is to be equally understood by managers, economists, and engineers;
3. The new forms should not introduce epistemological cuts.

A. S. C. Fernandes, *The Contribution of Technology to Added Value*,
DOI: 10.1007/978-1-4471-5001-5, © Springer-Verlag London 2013

Technology: Different Visions from Different Sciences

The etymology shows its origin in the Greek words *techne* and *logo*.The *techne* was referred to by Aristotle, in his book *Rhetoric* [3], as a method of systematizing knowledge. Today, quoting a contemporary American sociologist Merton [4], a technique consists of a standardized set of means, coupled to a particular purpose. It can therefore be a sequence of actions where either instruments or a rational sequence is used fulfilling certain criteria. The etymon *logo* describes the word, the study, and the knowledge. Accordingly, an upfront idea for the meaning of technology is the means and the ability for building something either an idea or a material object.

The term technology is modern and it was first used by the German philosopher Johann Beckmann, one of the first and most important scholars of industrial production, and a pioneer of scientific management. It appeared in the title of his books [5, 6]. Beckmann extended the older concept of technique to technology adding the notion of technical knowledge that is required to use a technique. Also, the term was used by Doctor and Botanist Bigelow [7], professor of mechanics at Harvard, MA, who published his lessons with the title *Elements of Technology* and explained the term as meaning the application of science to the arts and crafts. Finally, I point to the important reference of the *Massachusetts Institute of Technology*, founded in Cambridge, MA, in 1862. However, at the level of public discourse, the term only appears with regularity after the World War I, which made clear to the common citizen the terrible variety of the new technologies of war.

Three contemporary sociologists describe technology as a fuzzy term and difficult to contain in a single form [8]. They say that it is possible to identify three levels of meaning in the term technology: The physical object, an activity or a process, and the knowledge about the first two. They add that technology and society form a seamless web. This idea of technology, ambiguous and complex, has been treated alike by historians and sociologists, both Americans and Europeans. Merrit Roe Smith and Leo Marx [9] and also Hard and Jamison [10] edited two books with probably a deeper vision of the contemporary discussion on the essence and the role of technology. History and sociology understand the term technology, at its most abstract level, as what permeates human relations in society and in-between individuals and Nature. Science and technology are considered as cultural social constructs, a sort of a web that aggregates all members of society. Technology is thus described as an extraction residue of the relationship between people, their skills, and their activities.

In the context of Philosophy, according to Mack [11] and Kateb [12], the impact of technological awareness that emerged in the nineteenth century and early twentieth century forced to rethink traditional values, as it brought a new dimension to social relations. This new social component forced rapid and profound changes in human relations, which disturbed the culture, morals, ethics, and political systems. I present two examples with different social directions: (1) According to Jamison [13], the American philosopher John Dewey considered technology as a powerful contribution to the democracy; (2) in Germany, according to Zimmerman [14], the philosopher Martin Heidegger criticized the growing importance of technology and its influence on both contemporary culture and the political regime, for instance, supporting Hitler's national socialism and his rise to power. For Heidegger, technology was a product of modern metaphysics, in the sense that it developed before science. It was intrinsically linked to the concept of art and implied a thorough knowledge of material things. Also Spengler [15] considered technology as an instrument of power, predicting in his work the decline of the western civilization and that non-white civilizations would use our technology to destroy the West, which is an impressive speculation of the events of September 11, 2001 in New York.

Another important sociological reference is that of Karl Marx, whose analysis of social relations of production left an indelible mark in the centuries to come. In this field, he was not revolutionary because the technique (technology) was only taken as one element of the production system, dramatically centered on value created exclusively by workers and peasants. The technique (technology), he said, could be explained as a desideratum of Science [16, book I, vol. 1, part IV].

According to Feenberg [17], a philosopher of technology, there is in it an important and most relevant element: Its functionality. Technology performs a function in the economic transformation process. We can note that anthropologists and archaeologists characterize the culture of contemporary and ancient civilizations through their technology; and also that only the tangible part, as tools and housing, remains visible to our eyes, from what we may only imagine the necessary knowledge associated to them.

The concept of technology and its dependence on the social fabric is described in social sciences in a complex and extensive way. In its simplest form, it is possible to extract the three main levels in which they have an especial dimension:

- What is tangible, as tools, machinery, and systems;
- what is intangible, like general knowledge and specific skills to build them;
- and the functionality dimension.

Technology in a Primitive Society

Indeed, technology presents itself as a vehicle or as both a material and rational substrates of a social function. I shall illustrate, with a simple model of a primitive economy, the role of these three dimensions that seem to be part of the technology concepts: the tangible, the intangible and the functionality.

The simplest transformation process one can imagine involves only one individual and Nature; for example, the action of selecting and gathering a fruit from a tree. The individual, the active part of the process, acts according to his knowledge, in this case the need of food and information about that specific fruit. The product (output) of this process is a fruit ready to be eaten. The fruit on the tree or in the collector's hand shows no physical difference, but its status has changed throughout the process. At the beginning, it was part of Nature and in the end was owned by the individual and ready to be eaten. Between the initial and final states there is work, an action with a specific aim, which in turn has originated in the individual's knowledge. In this process, only knowledge and work play a role.

In the same context, let us consider now that the same individual selects and breaks a branch of a tree and prepares it in such a way that it could be used as a spear. This transformation process is similar to the previous but with a difference: The end product will not be consumed but preserved to be used later. Indeed, the individual plans to use the spear to reach and kill animals. In the hunting process, in addition to knowledge and work, there is a new element, the spear, which will have a relevant role in the process. This new element, which is not consumed in only one process even if it wears off, has the specific function of reinforcing the user's action and the effectiveness of his work. It should be noted that the spear is no longer a branch of a tree. Its shape and detail with which was prepared by its owner are such that a second individual will recognize, by examination, its purpose and function. The likely deduction of its function by simple inspection proves that there is knowledge embedded in the material, that it is no longer a branch of a tree but now a hunting spear.

This spear seems to have the three elements that we were looking for: (1) the tangible part—the substrate material; the intangible part—the embedded knowledge of both to use and to build it; and the function—reinforcing the individual's action. Is it a technological form or, simply, a technology?

Technology in the Modern World

In today's world, the diversity of technological forms is so great and their complexity so vast that basic elements, as its function or the idea of a tool, are no longer easily recognizable. However, the three attributes described earlier are still present and persist in all forms.

The intangible part of technology, the embodied knowledge, was divided into the knowledge necessary for its construction and the knowledge necessary for its use. In a primitive economy, it is possible that the two types of knowledge coincided in the same individual or social group. In today's world, however, they most probably belong to individuals who do not necessarily have a social relationship. The user is no longer the manufacturer or the creator and the latter does not act alone. From a basic tool, easily recognized, to a global computer system such as the Internet, recognizable only through a specific type of knowledge, go more than one million years and a few genetic steps of the Homo.

To classify the technologies of primitive societies is a trivial task because the division of labor was very simple. The earliest forms are certainly linked to hunting, gathering and war, followed by the fire related, domestic, and clothing. With the sedentary Neolithic came agricultural tools and irrigation, pastoralism, and house building. With the domestication of animals, slavery and the invention of the wheel appear those of transport and trade, in particular the ship, etc. Considering the materials, we would speak of stone or metal technologies, gold, copper, bronze, and iron; whereas the industrial sectors, those of railways, shipbuilding, construction, communication of information, etc. This is to make the point that the intangible part of technology, the embodied knowledge, is hold and retained in material forms for future social benefit. This accumulation of knowledge along the cultural evolution is, after all, the *acquis* of the civilizational memory, without which social evolution was impossible. Technological progress is the evolution of culture. However, we must remember that this knowledge, embodied in technological forms, is static and does not pass genetically to coming generations. The knowledge of an individual of this generation, accumulated and built over millions of years, must be apprehended in his short life, through his education. Such is the importance of technology and such is the importance of education.

Technology in Economics and Management

The accounting systems for companies and countries do not mention either technology or knowledge, but only capital and labor. Browsing the European System of Accounts [18], the word technology is absent, that is, those systems do not use the concept of technology, and therefore do not evaluate its importance much less its value.

In macroeconomics, the scenario is different because technology is everywhere, but the results are the same: Technology is neither objectively identified nor quantified. In fact, to comprehend how technology has contributed to economic growth has been a matter of central concern, especially after the World War II. During the 1960s and 1970s, the extraordinary growth of the United States of America, vis-à-vis other countries, led the Organisation for Economic Co-operation and Development (OECD) to investigate the causes of this difference. Macroeconomic growth models considered technology as an explicit parameter, models that were already discussed in detail in Chap. 2. The conclusion is that the concept of technology in macroeconomics is the sociological concept of technology, with the great ambiguity that we have seen above, what is incompatible with a quantification attitude. The notion of technology incorporates the notions of knowledge. The words technology, knowledge, technological knowledge, stock of knowledge, and technical change have been used as synonyms.

Over the past 50 years, the OECD has become the main world institution in what refers to the so-called Science and Technology System (S&T), to whom the European Union and Eurostat have joined recently. In all member countries, this S&T system monitors the spending on research and development (R&D), investment in knowledge, staff working in S&T, and some innovation parameters. This OECD system supports governments and their policies, allowing international comparisons of a large number of economic indicators, like the following: Staff mobility in R&D, publications and patents, and special cases of technological innovation. More recently, the data have been correlated with economic growth for many countries [19], validating the theoretical work and the underlying models. This system is based on a number of manuals [20–22] where, surprisingly, the concept of technology is not defined. What OECD has been measuring, in fact, is just the pace of innovation, especially innovation said technological, whatever this means.

In conclusion, also in macroeconomics the notion of technology, although widely used, is overlapped with knowledge, and not taken as a separate parameter, which could have in itself an influence on economic growth.

Engineering applies science to solve practical problems of society, and the solutions are technological forms. But engineering must be validated by management. Drucker [23, Chap. 1] said that *the manager is the dynamic, life-giving element in every business. Without its leadership the "resources of production" remain resources and never become production*. In the 1960s, production management became the most important part of corporate management. Even if economics and finances could not quantify technology, engineering knew that technology, however defined, was the product of their work. R&D became of maximum importance and during the 1980s the idea of production management was substituted by technology management. In the 1990s, the idea of technology was considered unsatisfactory because management realized that behind technology was people and their knowledge. This trend established the new knowledge management, where innovation has taken the role of the main and true growth engine. I will extend this analysis later when the concept of knowledge is to be reviewed.

Technology: The Operational Concept

Taking all attributes found and extracting those that are common to all areas of knowledge and persistent along the time, we get to the essential core of the concept. This list of attributes should define the new concept, and hopefully identify it against the concepts of knowledge and capital, once we have them identified as well. The fundamental attributes of technology are the following:

- It has a tangible form and is a produced good, and so is the result of a transformation process led by people and their knowledge.
- It is not a good to be consumed, rather to have a specific function in a production process.
- Its form is the result of the embodied human knowledge, what often suggests how it was done and how it should be used; but that knowledge is static, because it matches the knowledge with which it was developed.
- All the existing technologies contain all the knowledge that is socially available.

This definition of technology will only be complete when a rule is established to make a distinction with neighboring concepts like knowledge and capital. That will be done after the definitions of knowledge and capital operating concepts.

Knowledge: The Subject and the Object of Philosophy

The meaning of knowledge and how men come to know the truth is nothing less but the oldest demand, which has been the object of philosophical inquiry over the times. To be, to know, to think, and their relationship with the absolute appear in Philosophy as the faces of that primordial concern. The Parmenides' sentence *to be* or *not to be*, explained by Popper [24, Essay 3], and the Descartes' *cogito ergo sun* [25, II, art. 7] are singular proofs of that demand. Today, it is the object of science as well. Along the last century, it came into the fields of behavioral sciences, and more recently in scientific areas such as information processing and artificial intelligence. Moreover, Genetic Psychology, Neuropsychiatry, and Neurobiology came in good time into this field. Also, thermodynamics supports the idea that the organic life, such as the biological processes involving knowing, obeys the second principle. Finally, I must add that Management Science announced that the concept was also of its interest, and accordingly replaced, in the 1990s, the old personnel management by the new knowledge management.

To find out the most important attributes of the knowledge concept, let us start chronologically by Philosophy. The Western civilization builds on a dual principle, which divides the subject from the object, and considers the principle of causality as a rational imperative. How did the most relevant philosophical schools of thought explain knowledge and the process of knowing? They start from this dual principle, considering the individual as the subject and everything else as objects. Thus, knowledge is about how we understand what surrounds us, and the process of knowing is the way that outcome can be achieved. How has that process been understood?

In short, there were two main paths, set at the time of the classical Greek culture, paths that were brought together by Emmanuel Kant on the eighteenth century: The first considers that knowing is above all a rational individual effort, in accordance with a set of logical principles. It was established by Plato, explaining that knowledge is the process to get to the truth, process that is creative and rational. The truth being unreachable, the process is as important or more important than its purpose. The second, established by Aristotle, considers that knowledge is acquired above all by the interaction with Nature, through observation and systematization of the perceived. Again, the process of knowing is as important as the object. For Plato, the rational process is based on deduction, from the general to the singular; while for Aristotle, induction is the most fruitful method, generalizing from the particular.

This second path predominated for centuries, especially in the Middle Ages. Only humanism was to dethrone the dark ages and to mark the revival of the rational classical culture. Following the Renaissance, Descartes laid its foundations with its methodology of systematic doubt. The process of knowing was for him immaterial and built on innate ideas. In a way, he was returning to the Platonic conception of the primacy of reason. To this continental European rationalism was opposed the British empiricism, whose main figure was John Locke, arguing for both the supremacy of experience over the abstract reasoning and the absence of innate knowledge. It would be by experience that one could reach Nature, the source of all knowledge. The critical philosophy of Kant has grouped the two visions, rationalism and empiricism, showing their complementarity and synergic potential. Later, Hegel [26] asserted that more than the subject and the object, the greatest importance was on their dialectic relation. He sought to explain the engine of knowledge creation by the mediation of action or work, which characterizes the relationship between man and Nature.

Concluding, Popper [27] grouped knowledge into two types: subjective and objective. Subjective knowledge is tacit, intuitive, non-critical, consisting of dispositions and expectations. Objective knowledge, on the other hand, consists of the logical content of our theories, conjectures, and hypotheses [28, Chap. 3]. All objective knowledge constructs from the pre-existing knowledge, adding, adjusting, negating, or adding information via language or by observation, through conjectures and refutation. It is this interaction that makes objective knowledge to grow, a process where language, as a vehicle of critical reasoning, appears with a fundamental role. Published theories and discussions are examples of this type of knowledge.

From the described epistemology of knowledge emerges a set of three parameters, building the structure of the concept of knowledge:

- First, the two main centers—the subject, who are thinking humans; and the object which, in its simplest form, is everything else;
- Second, the interrelation between the subject and the object, which tends to be, in modernity, more important than the actual identification of the object;
- Third, a parameter that is only implicit in the discussion presented, the material support, biological or not; What are the main attributes of knowledge found along this analysis?
- It is part of the mind;
- It accumulates and builds on itself integrating the observation of Nature;
- It is fed by action, as language and work.

Knowledge, Data and Information

In the last century and up to the present day, scientific contributions have increased the understanding of this relationship between the mind and the environment, either natural or social. Neurosciences explain local activity at the level of brain cells with regard to physicochemical reactions. Also the live registration of brain activity, little by little, write the complete map of the brain regions and correlate them with all kinds of activities, including what is called feelings or creative action [29]. The different memories and consciousness are placed in preferential spatial areas. On the other hand, the language and its syntax, within the areas of information technology and software architectures, have contributed to the formulation of computer models whose functioning can hypothetically reflect the thinking process.

We described the structure of the knowledge concept and found some of its attributes. Still, its nature and its products have not yet been discussed, where we may find important attributes as well. I shall start with its nature, for the analysis of which the concepts of data and information will be used. These come from cybernetics and information theory.

Data are a number of bits and one bit is the unit of information. One bit has no qualitative meaning and only a dual quantity dimension, which is either, exists or does not exist. It is an abstract concept but it may be objectively defined, understood, and measured: a bit is a variation of anything, in its elemental form. It can also be the absence of variation, when the variation was possible or expected. In short, a bit is the elemental variation or no-variation in a binary system, 0 or 1. All physical reality can be described as data. A communication channel, by definition, supports data propagation, an example of which is the way we perceive information through our five senses. Data need a physical support either material or in any form of energy. In the context of the human brain, data may correspond to variations in specific chemical and physical processes, such as the potential difference of sodium or potassium ions between two sides of a cell membrane. The propagation of this potential difference along the dendrites is the way data flow within the brain. Thus, one can understand how neurons act as dynamic deposits of data and thus as memory.

Information is data arranged in a specific space–time structure, like for instance a specific sequence of bits. As data *per se* cannot have any meaning, it is the data structure that provides the qualitative part of information, just as two different sequences of the same musical notes make two different melodies. However, a sequence of bits can only constitute information if it is sensed by a receiver that can interpret its structure. As such, the emitter and the receiver must obey to the same communication rule; in other words, the receiver must be tuned with the emitter. Information theory and electronics have developed such communication systems and thus materializing these abstracts concepts like sensing, interpreting, and memorizing. Our brain possibly acts in a similar fashion: It detects data through the five senses and from the body itself; then, depending on being properly tuned, receives the corresponding information; finally, it may transform that information into knowledge. How is that done? What makes the difference between information and knowledge? There is a subtle but most important difference: The mind of the individual. There are situations when a set of information received by an individual is transformed into knowledge and situations when it is not. He may let it stay in memory and certainly decay with time or he may use it, relating it with other resident information, building new ideas, and applying them. The second alternative corresponds to the process of acquiring knowledge.

This explanation shows attributes that reinforce the conclusions of the last section. From the theory of information and from neurosciences, knowledge is a dynamic process not a static stage of any number of variables. Knowledge is a process that deals with present and incoming data and information, a biological dynamic progression that may have the assistance of instruments like rational rules, physical action or work, and the use of both a semantic and a syntax (in most cases a language). A simple quantity of data and information may be packed and stored or sent to any destination, as it has material substrate and corresponds to static concepts. On the contrary, knowledge, being a process, could not be packed and stored. From the process of knowing emerges a product, which may only be a rational entity but, with the support of work, will become an entity embodied or recorded in an out-of-the-mind material substrate. Then, that product of the knowing process will no longer be knowledge because it will become static. In brief, knowledge is a dynamic process involving data and information; it takes place within the mind as a biological process; and is supported by rational rules, language, and work.

Knowledge Management

It is a commonplace to say that society went out of the information age to enter the knowledge era. A small number of founders paved the way to this transformation. Tom Peters, from the beginning of the 1980s, insisted on the importance of people and their knowledge [30, Chap. 13]. Alvin Toffler and his Third Wave, which is after all the preponderance of knowledge, draws attention to this most important economic resource and warns that its value will grow, replacing the value of capital [31, Chap. 3]. Two new paradigms appeared: In the 1980s, the learning organization by Peter Senge [32]; and the organization that creates knowledge by Nonaka and Tacheuchi [33], in the 1990s. Also Peter Drucker [34], who is the author of the term knowledge worker and who said for the first time that "knowledge is the business" [35, Chap. 7], was also an important founder of the knowledge era.

What is knowledge for knowledge management? The concerns of knowledge management are focused on small social systems such as firms or groups of people with specific objectives, a hierarchical organizational structure and well-defined tasks. Knowledge management worries about creation, stocks, flows, distribution of knowledge, and the use of knowledge products.

There is a fundamental difference between individual knowledge and knowledge that is socially available, but the respective creation processes are often interrelated. Individual knowledge is acquired by a process in the individual's mind but, as we have seen, is supported by rational rules, language, and information inputs from the environment, natural and social. This is the typical education, training, and competence building process, which benefits the individual but rests with him/her. The objective of knowledge management is to arrange for the individuals to increase their knowledge and simultaneously to share it within the group such that it could be easily available and transformed in new products.

I shall review the ideas behind individual and business knowledge as seen by knowledge management, enriching the list of attributes of this concept. Davenport and Prusak [36] focus on the fact that, in an organization, knowledge becomes embedded in objects, documents, routines, processes, standards, etc. In the individual, on the other hand, knowledge is described as incorporating information and experiences. Alle [37] deals specially with the practical aspects of knowledge management, criticizing and suggesting rules and maps for various situations, and largely ignoring the individual creation of knowledge. Leonard [38] writes about how to manage the knowledge assets of a company, recreating and developing them. Based on the real cases, it confirms that knowledge management aims at the best possible combinations between the individual knowledge of each employee and the firm's activities in which she is involved. It is the appropriate management of this combination that promotes innovation. She refers to skills and knowledge as being incorporated into materials or residing in people's heads. For the latter case, she uses the term proprietary knowledge, emphasizing the idea that there will be a part that belongs exclusively to the individual.

Companies need to know how much knowledge they have, where it is and how it could be expanded. Moreover, they need to value it and to know how it contributes to value adding. The knowledge of a firm, putting aside the employee's knowledge, match up only to its knowledge products. Boisot [39], who wrote one of the reference books on knowledge assets or stocks of knowledge, states that they have tangible characteristics, like socio-physical systems or technology, and intangible characteristics, like skills. He is clearly wrong when saying that knowledge assets, in contrast to physical assets, could last forever. In fact, knowledge, either objective or subjective, in the human mind or embedded in any material substrate, decays, or depreciates with time.

Malhotra [40], in his excellent review on the methods of valuing knowledge, uses the term capital, namely intellectual capital. This term was launched with international projection by Edvinsson [41, 42, 43], who defined it in 1991 as a combination of human capital and structural capital. The latter would include aspects such as capital related to customers and suppliers, intellectual property, and other intangible items. A similar term, human capital, would be, in the opinion of Sullivan [43, 44], the sum of the tacit knowledge of a firm's employees.

It is very clear that knowledge management often overlaps the concepts of knowledge, technology, and capital, though recognizing the fundamental differences between the individual knowledge and the products of knowledge.

Knowledge: The Operational Concept

Following this analysis, I propose that the operational concept of knowledge we intend to rebuild is characterized by the following main attributes:

- It is a dynamic process concerning the subject and the objects, taking place in the biological mind;
- It has a tacit and intuitive component, whose evolution is very difficult to establish;
- It has a rational component which, by the interaction of intuitive processes with external environment information, enables the reorganization of data and information in memory by creating more and new meanings;

- It involves the dynamics of rational and intuitive components and is supported by language and expressed by action and work;
- It has as final products static forms of knowledge embedded in material substrates. Thus, work and its products appear as visible forms of human knowledge, although in a static form.

As stated for the case of technology, the definition for the operational concept of knowledge will only be complete when a rule is established to make a distinction with neighboring concepts like technology and capital. That will be done after the analysis of the capital concept.

Capital: From a Primitive Society to the Middle Ages

The word capital is derived from the Latin *caput*, meaning head, and was introduced in the English language by the thirteenth century. The common idea of capital is associated with wealth, which approximately matches its meaning in Economics. Currently, the term is applied, with a suitable adjective, to almost anything from which one can take some kind of economic advantage, and I shall use this general significance to start the search for its fundamental attributes. We will look first into a primitive society, where it is clearer to identify the most important production factors.

When the concept of technology was analyzed, a spear used for hunting was identified as a technology form. Roughly, it could be a form of capital as well. The idea of storing salted meat for the survival of a small community may also suggest capital accumulation. In this case, it may be used for future consumption or for exchange with other goods. Another example, perhaps the most appropriate to understand how it may be different from technology, is the case of the surplus of cereal, which appeared in agrarian societies some 10,000 years ago. This surplus is, from the total harvest, what is not consumed. However, it is more than a simple surplus: To store and treasure the best grain is vital to sustain the agrarian activity. The stored grain as seed for next year's sowing is perhaps the first clear form of capital. It is a produced good, but it is not to be consumed and it is indispensable for the production process, a par with land. It has the property of reproducing more grain and thus more wealth. As technology, it has a function in the production process. Sedentary populations would build permanent shelters, for themselves, for cattle and for storing surpluses. Those houses would also play a role in the production process. Which role? Are they technological forms or capital forms? Which are indispensable to a greater degree?

With increasing surplus and savings, barter and commerce emerged as a vital activity for settled populations, a need that brought together producers and users and the emergence of market places. Later, long distance commerce asked for special goods that would be easily exchanged for most of the merchandise, goods with a high value density and worldly acknowledged as valuable. Cattle, wheat, salt, sea shells, and precious metals were used for that purpose. The value of these exchange goods was liquid, in the sense that it would take the "form" of the value of any other good. This is how the idea of money or currency was born, a produced good with an intrinsic value that could flow between producers, users, and merchants without being consumed. As it became increasingly important for production and commerce, money rose to the status of a principal resource or capital (what comes ahead), and with it banking, guaranties, and credit turned out to be more easy and operational. What were the resources that would come ahead, and so act as capital? In general, all forms of wealth could act as capital, like money, land and slaves, and up to a certain point credit.

In the Western world, from the older civilizations to classical Greece and Rome, commerce developed rapidly. The importance of trade was facilitated and fueled by the progressive use of coins, introduced in Greece perhaps around the seventh century BC. However, agricultural activity was highly predominant over the commercial, industrial, and artisanal, at least until the eighteenth century [45, Chap. 2]. The gross national product (GNP) per capita in the world has increased little or nothing in the first millennium of our era and just grew 50 % between the year 1,000 and 1,820 [46, Table 1–2]. From the fourth to the seventh centuries, Southern Europe was invaded by peoples of the North and Eastern regions, replacing the *pax romana* by their barbaric and illiterate cultures. In the eighth century, the Muslims seized political power in most of Iberia and Sicily, as well as North Africa and the Eastern Mediterranean. From the eighth to the eleventh centuries, the Vikings moved South by sea, settled permanently in England, the Netherlands and Normandy, entering the Mediterranean up to Sicily, paving the way for new lines of trade. At the end of this problematic period, the first reappearance of important wealth creation and capital accumulation lies in the Venetian Republic, twelfth to the fourteenth century, a center of extraordinary trade reputation and dimension, who retained permanent links with the East through profound political changes in Europe and the Middle East. Currency, banks, and insurance companies had an incomparable growth, only overtaken by the Dutch empire in the seventeenth century.

Capital: From Mercantilism to Liberalism and Marginalism

The Portuguese and Spanish discoveries of the sixteenth century opened the door to new worlds and created viable alternatives for trade with the East and West. The amount of precious metal that entered Europe was extraordinary, contributing in a substantial way to new capital accumulation.[1] Currency, gold, and silver were the best recognized forms of wealth and the forms most clearly identified with capital. This characterizes the mercantilist idea of capital; however, the reality from the seventeenth century onwards pointed to other new and most relevant dimensions.

[1] Between 1,500 and 1,800, one thousand and seven hundred tons of gold and more than seventy thousand tons of silver travelled from America to Europe (Maddison [46], Table 2–8).

The development of new empires implied more than just available currency. For example, the naval art was not a product that could be purchased just like any other good. Was this art a new form of capital? And the huge fleets, property of numerous commercial companies? Were those ships also a form of capital as important as currency? And the companies' headquarters, customs, ports, and palaces were them forms of capital as well? To what extent could the knowledge to build and manage those commercial systems and the assets themselves, ships, headquarters of companies, roads, palaces, and slaves be used to reproduce and create more wealth?

It was during the mercantilist times that the difference between wealth and source of wealth was better understood. Wealth could be used for consumption or as investment to hopefully result in more wealth. As such, the idea of capital became tightly associated with investment, separately from the notion of wealth. Moreover, countries recognized that it was not only capital in currency forms and the extent of land that made the difference, as William Petty explained to his King, comparing the economies of England, France, Holland, and Zeeland, that a small country with few men can be as rich as a large country [47, Chap. 1].

The main face of physiocracy, Francois Quesnay, modeled for the first time an agrarian production process, considering land as the sole source of value [48]. His system divided the capital that a farmer needs in two parts, fixed and circulating. At the same time, in Scotland, Smith was the first economist to defend the virtues of the market economy. He considered that the assets of a country could be divided into three parts: Consumption and capital, the latter separated into fixed capital and circulating capital [49], (book II, Chap. 1). Fixed capital was defined, once again, as goods not to be consumed, i.e., goods that would be used in the production of future goods. As Quesnay and Smith, Ricardo adopts the ideas of fixed capital and circulating capital and exemplifies them both in productive systems of his time and in primitive economies [50, Chap. 2]. As the names suggest, fixed capital refers to machinery, instruments, buildings, and even workers' skills. The circulating capital included credit and money, goods in transit, and products not yet sold. Smith, Ricardo, and Jean Baptiste Say are the three founders of classical economics. From the analysis of their contributions to the capital concept, I conclude that it was enlarged substantially. From the seed, in a primitive economy, it was added in Antiquity currency and credit, and at their time tools and machinery, raw materials, merchandise, goods produced but not yet sold, and even the workers' knowledge and skills.

At the time of the last editions of Marx's The Capital, three men developed a new economic vision: Stanley Jevons in England, Carl Menger in Vienna, and Léon Walras in Lausanne. Jevons said that capital was nothing more than a form of the workers maintenance [51, paragraph 21]. They founded what today is called the marginalism school. The main contribution was to the theory of value; more precisely, they explained the mechanism of price formation using mainly the buyer's decision instead of the production costs. This new perspective was applied to the capital itself, whether as a good, or in the form of currency and credit. This resulted in an important clarification between the interest and the profit, but did not bring important changes to the concept of capital. This was made clear in the review of the concept of capital by Irving Fisher [52], where he discusses the meanings of capital, stock, and surplus, words in vogue at the time, to conclude that they refer to approximately the same realities. As another example, Alfred Marshall, in his Principles [53, book 2, Chap. 4], reviews the different capital designations and its characteristics, from Smith to Ricardo and Stuart Mill, and concludes that fixed capital and circulating capital are already a consensual reference.

Human and Social Capital

Above, when the concept of knowledge was analyzed, the notion of business knowledge was reviewed as well as the modern ideas of intellectual capital and human capital. Earlier, Ricardo had mentioned worker skills as a capital good. Also, slave work has been referred to as a form of capital. Therefore, the idea of human capital has been around for more than a century. More recently, Schultz [54] established the contemporary base for considering people's knowledge as a form of capital, human capital.

Ever since the Schumpeter analysis on the importance of innovation and creative destruction, which he emphasized that characterized capitalism [55, Chap. 7], firms and corporations gave additional attention to the human element, and the role of workers was re-evaluated, in particular the knowledge workers. Schumpeter believed that the idea of capital could be split into the foundation capital and working capital [56, Chap. 3], but he favored the narrower definition of capital as means of payment, clearly different from forms of capital like machines and buildings. Nevertheless, he mentions knowledge and projects, which constitute the core of new business initiatives and production processes, as forms of capital.

What today is known as human capital has many forms but can be reduced to its essential—it directly depends on human knowledge. It is true that routine and artless work also originates from the mind and human knowledge. However, it was realized the higher importance of knowledge as a factor before and in parallel with work. This is the line along which Boisot defines knowledge goods as knowledge that can return benefits over time, which is equivalent to the role that is attributed to capital [39, Chaps. 2 and 7].

Other kinds of capital, such as intellectual capital, have been proposed. Stewart, another founder of this concept, defined it, within a firm, as the summation of all human knowledge relevant to its competitive advantage. That knowledge was not only the one embedded in files, books, and patents, but also the talent and expertise of employees [57, Foreword and Chap. 4]. In this context, the management of intellectual capital relates to the adaptation of tools and machinery to the human way of thinking and acting, what is an opposite attitude to the one of the beginning of the fourteenth century when management claimed to adapt workers to machines.

Social capital is a term even more abstract. A recent paper characterizes well the attributes of this form of capital [58]: Trust and civic standards, cultural values as altruism and tolerance, social values as institutions of social relations. The educational system itself and the judicial system of a country appear as forms of social capital at macro level, or, at the micro level, a permanent training program in a company, for example. There are also recent economic studies that correlate the levels of this type of capital with economic growth in OECD countries [59, p. 57]. In this study, trust among economic agents is studied, bringing this concept, formerly a sociological notion, to the category of social capital.

Capital: The Operational Concept

Considering all the attributes found and taking those that are common to all areas of knowledge and persistent along the time, the fundamental attributes of capital were found to be the following:

- It has a tangible form and is a produced good or a natural resource;
- It distinguishes clearly from a consumer product;
- It has a role in economic production processes;
- It is a form of value accumulation and so reproduces, accumulates, carries in itself static forms of information, knowledge, and work;
- It is easily transformable in form and has a wide applicability.

As for the cases of technology and knowledge, the definition for the operational concept of capital will only be complete when a rule is established to make a distinction with neighboring concepts like technology and knowledge. That will be done for the three concepts in the next section.

Knowledge, Technology and Their Border Line

The most important conclusion that was found in this analysis is that knowledge is the basis and the first cause, and therefore technology and capital depend on it and somehow reflect it. But there are important and most significant differences. Knowledge is a dynamic process and a technology form is tangible and measurable. In fact, a specific type of knowledge and the corresponding technological form may contain approximately the same information. However, while knowledge, being a dynamic and subject intrinsic process, is continually evolving; on the other hand, a technological form embodies a concrete and objective set of information in a static form, i.e., without evolution.

A technological form, at a given time, represents a crystallization of individual knowledge, which will remain the same along the time embedded in the respective material form. If technological forms are static, so is technology, understood as a set of technological forms. This does not exclude, though, that technology, seen as a succession of technological forms deteriorating in time and being replaced by others perhaps better suited and in higher quantity, cannot be interpreted as a dynamic social reality. But this dynamic technological evolution, as the expression of technological culture, is already a social non-operational concept.

It is so decided criteria for determining whether we are dealing with a form of knowledge or technology. The criteria are to be or not to be dynamic, and to be subjective or objective. Knowledge is subjective and a dynamic process, and a technological form is a concrete reality, objective, and static. This is the border line between the operational concepts of knowledge and technology.

Technology, Capital and Their Border Line

The border line between technology and capital is more challenging to establish. Reading again their main attributes, and analyzing one by one, we see that the concepts of capital and technology are almost equivalent. The only two significant differences are: (1) The wider applicability of capital forms in contrast with a more rigid applicability of technological forms; (2) the possibility of a natural resource being a form of capital, like land, what cannot happen to technology.

A few examples may illustrate how these differences can be found. A simple technological form like a spear for hunting, or a complex form such as a screw machine, or even an abstract and sophisticated form such as an information system have well-defined purposes. In other words, the spear cannot make screws and an information system cannot be used for hunting. As for capital forms, they should have flexibility of application, that is, a particular form of capital should serve for different operational functions. For example: (a) A firm owns the building in which it operates. This building is an asset and serves a wide variety of functions, since it can house different industries, services of one type or another, assembly lines or offices. (b) The firm's money and credit are current assets with a very flexible application, as they can be used to pay salaries, buy intermediate products, pay interest on loans, acquire patents, etc. In brief, according to this criterion, the building and current assets are essentially forms of capital and they are definitely not technology forms. As for a natural resource, as land, the sea, the sun, or the wind, there use is reflected differently in the list of firms' costs. Land, if and when used, has always been considered a production factor, and today a capital asset subject to depreciation. Fish farms could be handled in the same way. In both cases, they represent forms of capital and not technological forms. Alternatively, as in the most cases of the mining industry, the use of land is subjected to an annual fee, a sort of a rent. In this case, the respective cost is just a service provided by third parties.

The border line that can be used to discriminate between forms of capital and forms of technology is drawn analyzing the applicability of the respective asset and deciding in each particular case what is technology and what is capital. Most of the times, it is not clear and thus the proposed algorithm will make it possible to a specific asset to be classified in both types simultaneously, a percentage technology and a percentage capital. That would need an objective criterion, what will be explained in detail in the last section of Chap. 3.

References

1. Fernandes ASC (2007) Mais Conhecimento e Tecnologia para Desenvolver a Economia Portuguesa, Lisboa: Fundação Calouste Gulbenkian
2. Fernandes ASC (2012) Assessing the technology contribution to value added. Technol Forecast Social Change 79:281–297
3. Aristóteles Rehtoric (2012) Book 1. http://classics.mit.edu/Aristotle/rhetoric.1.i.html. Accessed April 2012
4. Merton RK (1973) The sociology of science: theoretical and empirical investigations. The University of Chicago Press, Chicago
5. Beckmann J (1779) Beyträge zur Oekonomie, Technologie, Polizey und Cameralwissenschaft, Göttingen. http://www.thur.de/philo/beck2.htm. Accessed April 2012
6. Beckmann J (1802) Anleitung zur Technologie oder zur Kenntniß der Handwerke, Fabriken und Manufacturen, Göttingen. http://www.thur.de/philo/beck2.htm. Accessed April 2012
7. Bigelow J (1831) Elements of technology, Hilliard, cray, little and Wilkins: Boston (1ª edn 1829). http://ia600604.us.archive.org/18/items/elementsoftechno01bige/elementsoftechno01bige.pdf. Also in http://books.google.pt/books?id=ed8JAAAAIAAJ&printsec=frontcover&hl=pt-PT#v=onepage&q&f=false. Accessed Mar 2012
8. Bijker WE, Hughes TP, Pinch T (eds) (1987) The social construction of technological systems. MIT Press, Cambridge
9. Smith MR, Marx L (eds) (1994) Does Technology Drive History? The Dilemma of Technological Determinism. MIT Press, Massachusetts
10. Hard M, Jamison A (eds) (1998) The intellectual appropriation of technology. MIT Press, Cambridge
11. Mack A (ed) (1997) Technology and the rest of culture. State University Press, Ohio
12. Kateb G (1997) Technology and philosophy. In: Mack A
13. Jamison A (1998) American anxieties: technology and the reshaping of republican values. In: Hard M, Jamison A (eds) The intellectual appropriation of technology. MIT Press, Cambridge
14. Zimmerman ME (1990) Confronto de Heidegger com a Modernidade. Lisboa: Instituto Piaget. Portuguese translation from Heidegger's Confrontation with Modernity: Technology, Politics, and Art, Bloomington: Indiana University Press
15. Spengler O (1926 and 1928). The decline of the west (Vol. 1, "Form and Actuality"; Vol. 2, "Perspectives of World History"). New York: Alfred A. Knopf. Comments and parts fo the text in http://www.fordham.edu/halsall/mod/modsbook.html. Accessed Mar 2005
16. Marx K (1887) Capital. First English edition. http://www.marxists.org/archive/marx/works/1867-c1/. Accessed April 2012
17. Feenberg A (2003) From Essentialism to Constructivism: philosophy of technology at the crossroads. In: Higgs E et al Technology and the good life. UCP, Chicago
18. ESA95 (1996) European system of accounts. Office for official publications of the European communities, Luxembourg
19. Bassanini A, Scarpetta S (2001/2) The Driving Forces of Economic Growth: Panel Data Evidence for the OECD Countries. OECD Economic Studies, N° 33
20. OECD (1994) Proposed standard practice for surveys of research and experimental development, 5a edn. Frascati Manual, Paris
21. OECD and Eurostat (1995) Manual on the measurement of human resources devoted to S&T. Canberra Manual, Paris

22. OECD and Eurostat (1997) Proposed guidelines for collecting and interpreting technological innovation data: Oslo Manual (2a edn). Paris
23. Drucker P (1954) The practice of management. Harper & Brothers Publishers, New York
24. Popper KR (2001) The world of parmenides. Routledge, Londres
25. Descartes R (1971) Princípios da Filosofia. Lisboa: Guimarães Editores. Portuguese translation from French edition of 1647. http://www.literature.org/authors/descartes-rene/reason-discourse/. Accessed April 2012
26. Hegel (1955) La raison dans l'histoire. Union Générale dÉditions, Paris
27. Popper KR (1996) O Conhecimento e o Problema do Corpo-Mente. Lisboa: Edições 70. Portuguese translation of Knowledge and the Body-Mind Problem
28. Popper KR (1979) Objective knowledge. Oxford University Press, Oxford (Revised edition)
29. Damásio AR (1999) How the brain creates the mind. Scientific America, Dec 99, end of millennium special issue (74–79)
30. Peters TJ, Austin NK (1985) A Paixão pela Excelência. Lisboa: Pensamento Editores Livreiros (Portuguese translation from A Passion for Excellence)
31. Toffler A, Toffler H (1995) Criando uma Nova Civilização. Lisboa: Livros do Brasil (Portuguese translation from Creating a New Civilization)
32. Senge PM (1990) The fifth discipline: the age and practice of the learning organisation. Century Business, London
33. Nonaka I, Tacheuchi H (1995) The knowledge creating company. Oxford University Press, Oxford
34. Drucker P (1993) Post-capitalist society. Harper-Collins, New York
35. Drucker P (1964) Managing for results. Harper & Row Publishers, New York
36. Davenport TH, Prusak L (1998) Working knowledge. HBS Press, Harvard
37. Alle V (1997) The knowledge evolution. Butterworth Heinemann, Newton
38. Leonard D (1998) Wellsprings of knowledge. Harvard Business School Press, Boston
39. Boisot MH (1998) Knowledge assets. Oxford University Press, New York
40. Malhotra Y (2003) Measuring knowledge assets of a nation: knowledge systems for development. Invited presentation in Ad Hoc group of experts meeting knowledge systems for development, United Nations Headquarters, New York
41. Edvinsson L, Frij A (1998) Skandia: three generations of intellectual capital. In: Imparato N (ed) Capital for our time: the economic, legal, and management challenges of intellectual capital. Hoover Institution Press, Stanford, pp 192–201
42. Edvinsson L, Malone MS (1997) Intellectual capital: realizing your company's true value by finding its hidden roots. Harper Business, New York
43. Imparato N (ed) (1999) Capital for our time. Hoover Institution Press, Stanford
44. Sullivan PH (1999) Extracting profits from intellectual capital: policy and practice. In: Imparato N (ed) Capital for our time: the economic, legal and management challenges of intellectual capital. Hoover Institute Press, Stanford
45. Braudel F (1985) A Dinâmica do Capitalismo. Lisboa: Teorema. Portuguese translation from La Dynamique du Capitalisme, Éditions Arthaud, Paris, 1985
46. Maddison A (2001) The world economy. OECD, Paris
47. Petty W (1690) Political arithmetick. London. http://socserv2.socsci.mcmaster.ca/~econ/ugcm/3ll3/petty/t. Accessed April 2012
48. Quesnay F (1985) Quadro Económico. Lisboa: Fundação Calouste Gulbenkian, 3rd edn. First published in 1758
49. Smith A (1956) Inquiry into the nature and causes of the wealth of nations. Collier & Son Corporation (original published in 1776), New York
50. Ricardo D (1821) On the principles of political economy and taxation (3ª edn). http://www.econlib.org/library/Ricardo/ricP4.html. Accessed Sept 2012
51. Jevons S (1866) Brief Account of a General Mathematical Theory of Political Economy. J Royal Statistical Society, London. http://www.marxists.org/reference/subject/economics/jevons/mathem.htm. Accessed February 2013
52. Fisher E (1904) Precedents for defining capital. Q J Econo 18:386–408 http://socserv2.socsci.mcmaster.ca/%7eecon/ugcm/3ll3/fisher/capital4. Accessed June 2004

53. Marshall A (1890) The Principles of Economics. http://www.econlib.org/library/Marshall/marP.html. Accessed February 2013
54. Schultz TW (1961) Investment in human capital. Am Econ Rev 51:1–17
55. Schumpeter J (1976) Capitalism, socialism and democracy. New print 1992 of the 5th edn of London, Routledge (1st edn in 1943)
56. Schumpeter J (1983) The theory of economic development. Transactions Publishers (1997), London
57. Stewart TA (1997) Intellectual capital. Doubleday, New York
58. Chou YK (2006) Three simple models of social capital and economic growth. J Socio-Econ 35(5):889–912
59. Temple J (2001) Growth effects of education and social capital in the OECD countries. OECD Economic Studies 33 2001/2

Printed in the United States
By Bookmasters